COMPUTER INTERFACING:

A Practical Approach to Data Acquisition and Control

William H. Rigby
Professor and Department Head
Electronics Department
Industrial Technologies Department
Northern Michigan University

Terry Dalby
Field Service Engineer
American Microwave Corporation
Adjunct Instructor
Northern Michigan University

Prentice Hall
Englewood Cliffs, New Jersey 07632

Library of Congress Cataloging-in-Publication Data

Rigby, William H.   Computer interfacing : a practical approach to data acquisition
and control / by William H. Rigby, Terry L. Dalby

   p. cm.

   Includes index.   ISBN 0–13–288374–0

   1. Computer interfaces.   2. Automatic data collection systems.

3. Microcomputers.   I. Dalby, Terry L.   II. Title.

TK7887.5.R54   1994

006—dc20   94–14132
                                          CIP

Acquisition Editor: Holly Hodder
Production Editors: Rose Kernan and Fred Dahl
Designer: Fred Dahl
Cover Designer: Marianne Frasco
Project Production Manager: Ilene Sanford

© 1995 Prentice-Hall, Inc.
A Simon & Schuster Company
Englewood Cliffs, New Jersey 07632

No patent liability is assumed with respect to the use of information contained herein. While
every effort has been taken in the preparation of this book, the authors assume no responsibility for errors or omissions. Neither is any liability assumed for damages resulting from
the use of the information contained herein.

Information contained in this product should not be used in life support or critical applications.

IBM, IBM PC, XT, AT, PS/2, and Micro Channel Architecture are registered trademarks of
International Business Machines Corporation.
Labtech Notebook–Labtech, Laboratory Technologies Corp.
Unkelscope and Unkelscope Junior–Unkel Software, Inc.
BASIC–Dartmouth College
Intel–Intel Corporation
Turbo C–Borland International

Printed in the United States of America
10 9 8 7 7 6 6 5 5 4 3 2

ISBN 0-13-288374-0

Prentice-Hall International (UK) Limited, London
Prentice-Hall of Australia Pty. Limited, Sydney
Prentice-Hall Canada Inc., Toronto
Prentice-Hall Hispanoamericana, S.A., Mexico
Prentice-Hall of India Private Limited, New Delhi
Prentice-Hall of Japan, Inc., Tokyo
Simon & Schuster Asia Pte. Ltd., Singapore
Editora Prentice-Hall do Brasil, Ltda., Rio de Janeiro

To our families:
Sue, Craig, and Susan—W. H. Rigby
Ruth, Erin, Sean, and Ryan—T. L. Dalby

CONTENTS

PREFACE

In recent years the disciplines of electronics and computers have become so inter-twined that it is difficult to differentiate between what were two separate disciplines. Not long ago, students of electronics studying their chosen field focused almost exclusively on applications of electronics including electrical circuits, radio com-munications, television circuits, and similar applications for electronic circuits. While the electronics students were studying transistor amplifiers and Colpitts Oscillators, other students majoring in computer science were writing software, wiring Plug-Boards, and sitting at Key Punch Machines punching holes in "IBM" cards.

The introduction of the personal computer (PC) in the late 1970s changed forever any separation that existed between these two fields of study. Because of the proliferation of PC s the fields of electronics and computers would be intertwined. The low cost of PC s provided computers for "the masses" and opened a new area of study for students of electronics. Today the intertwining of computers and electronics is so close that these two fields of study have become one. Computer science curriculums have taken on some of the character of electronics programs and electronics programs are teaching many topics previously taught by computer science programs. A visitor to the electronics laboratory in any college, technical school, or high school vocational program will find computers in nearly all of the laboratories and classrooms. In the few years since the introduction of the PC, areas of study such as the teaching of programming languages, the use of application software, applications of computer I/O, and the topic of computer maintenance have all become recent additions to the body of knowledge taught in contemporary electronics programs.

Because of these changes this text will provide a resource for students who want to apply personal computers to applications involving interfacing, digital I/O, analog I/O, data acquisition, and computer control of external electrical devices such as pumps, fans, and other electrical devices.

The book will provide a balance between theory and practice as both are applied to the fields of computer interfacing, programming, interface wiring techniques, and the fundamental knowledge of computer equipment that is necessary to carry out computer control.

This text focuses on the use of standard interfacing boards, which are produced by several manufacturers. For purposes of clarity the laboratory exercises described in this book use interfacing boards produced by Keithley MetraByte Corp, 440 Myles Standish Blvd, Taunton, MA 02780. The following I/O boards are needed to complete all of the exercises in this text; PIO-12, DAS-4, DAC-02, and CTM-05. These boards are inexpensive and offer colleges, technical schools, and other educational programs a low cost way of buying industrial standard interfacing boards at a reasonable cost. The specified I/O boards are compatible with the PC/XT/AT bus.

Examples of the programming instructions necessary to implement the above products have been provided throughout the book. Brief programming examples (program kernels) are presented in both BASIC and C.

These two programming languages were selected for inclusion in this book based on several factors. The prime factor considered is the long standing popularity of the BASIC language and the increasing popularity of the C language. Students and faculty may choose to use either or both languages in the Lab Exercises. Always, the programming examples have been kept straight forward in an attempt to avoid "muddying the water" with obscure programming techniques or the use of programmers "tricks."

The use of PCs in I/O applications, data acquisition, and control of electrical and mechanical equipment is a most fascinating field that provides a high level of satisfaction to the student of electronics. Enjoy as you learn!

William H. Rigby
Terry L. Dalby

Marquette, MI

ACKNOWLEDGMENTS

The authors would like to thank all of the individuals who have helped with the development of this text. Thanks is extended to the students enrolled in ET 410 *Data Acquisition and Control* at Northern Michigan University for their suggestions and patience as drafts of the text were field tested.

A text of this type, one that focuses on commercial or industrial products, would not be possible without the help of equipment manufacturers and vendors. Appreciation is extended to the hardware and software manufacturers, and distributors, who provided advice, information, and copyright permission during the preparation of this text. The authors would like to single out the following individuals for recognition.

Mary Scott & William Unkel, Unkel Software, Inc.
Darci Parks, Industrial Computer Source
Karim Sidi & Robert Molloy, CyberResearch, Inc.
Markey Mason, Intel Corporation
Christine Mauceri, Laboratory Technologies, Inc.
Nayan Pradhan, Contec Microelectronics U.S.A. Inc.
Sharon Strong & Maria Psychopedas, Omega Engineering, Inc.
Steve Behar, Thermometrics, Inc.
Robert Auman, Intelligent Instrumentation
Steve Fox, Elmwood Sensors, Inc.
Fred Jones, SenSym, Inc.

Ann Marie Flynn, Data Translation
Christy Murphy, Micro Switch
Steve Chase, Gordos Corporation
Robert Ferran, Crydom Company
Ben Bailey, Computer Boards, Inc.

Special recognition and a sincere **Thank you** is extended to Jack Chase, **Marketing Manager**, Keithley MetraByte Corporation. Jack has been supportive of this project from the beginning. As the concept of the text was developed, through the writing of the last chapter, Jack Chase has provided assistance in the form of suggestions, product literature, and most of all encouragement.

CHAPTER **1**

An Overview of Data Acquisition & Control

DEFINITION OF DATA ACQUISITION & CONTROL

Data acquisition and control may take a variety of forms. At the simplest level, data acquisition can be accomplished by a person, using paper and pencil, recording readings from a volt meter, ohmmeter, or other instrument. For some situations this form of data acquisition may be adequate. Many applications, especially situations where the number of data points being recorded is relatively low and the frequency of the readings is slow enough for humans to record the data, this "low-tech" method of recording data is very appropriate. Those of us interested in technology, and applying technology to accomplish tasks, need to be sensitive to the suggestion that we tend to "complicate life" and often use technology solely for the sake of using technology.

However, data recording applications that require large numbers of data readings or situations where very frequent readings are necessary, must rely on instruments or computers to acquire and record the data. These situations may require the use of strip chart recorders, data loggers, or personal computers to record large amounts of data, or to record data occurring at fast rates.

Control of electrical devices can also occur at various levels of sophistication. At the "low tech" level, a simple electrical switch can control a motor, conveyor belt, or other electrical/mechanical device. Often this approach to technology is adequate and frequently preferred over more complicated methods of control, particularly where the cost of a sophisticated control system is not justified. Again, we do not want to use "high tech" where high tech is not justified! (See Figure 1.1.)

Other control applications may require a more sophisticated approach. Industrial equipment including robots, integrated process control systems, and conveyor belts may need the processing power and flexible programming ability of a personal computer to provide the control needed in complicated systems.

Personal computers are being used in data acquisition and control applications along with special data acquisition & control (DA & C) boards installed in the computer. These boards are inserted into the chassis of the computer and allow the computer to be connected to sensors, probes, and other monitoring devices much like a voltmeter is connected by the probes to the voltage source being measured.

These special data acquisition & control (DA & C) boards also allow the computer to control (turn ON or OFF) external electrical devices, much like a switch controls an electrical load. Thus, through the use of DA & C boards the computer can become an integral part of a comprehensive and very flexible data acquisition & control system. The computer provides many attractive features to the data acquisition and control system such as high clock speed, programming flexibility, mass data storage, low cost, high computational power, and a level of sophistication not available to discrete circuits. The inclusion of DA & C boards in a computer allows the technician or engineer to connect the computer directly to remote electrical sensors that will provide electrical measurements to the computer and to electrical devices such as relays that allow the computer to control these electrical loads. (See Figure 1.2.)

Data acquisition & control boards provide new and expanded applications for computers in industry, research, and any application where process monitoring, control, or data acquisition needs to be performed. Most computer users are familiar with the traditional applications of computers such as word processing, data entry,

Figure 1.1 Personal Computer and Data Acquisition and Control Boards
(Courtesy Intelligent Instrumentation)

Figure 1.2 Personal Computer and DA & C Board
(Courtesy Data Translation)

or other uses in which all communication with the computer is via the keyboard. This type of input and output may be thought of as using the "front door" of the computer. This is typical of traditional computer applications. Through DA & C boards, the technician or engineer can access the computer in a non-traditional way. Access through the DA & C board can be visualized as getting to the computer via the "back door." Naturally programming the computer and manual control of the computer is still accomplished by the keyboard. The "back door" simply provides another dimension for getting data into the computer and using the processing and decision-making power of the microprocessor. Using the input/output (I/O) capabilities of a computer creates a new body of knowledge for technicians and engineers.

The body of knowledge required of today's technicians and engineers includes what previously had been the purview of computer gurus along with the body of knowledge traditionally associated with the field of electronics. The electronics technician today is concerned with computer programming, developing software algorithms, interfacing, signal conditioning, as well as operational amplifiers, transistors, electric motors, and an endless variety of digital integrated circuits. Therefore, the body of knowledge for electronics technicians and engineers has vastly increased during the past few years. While this expansion may present difficulties and hurdles, it has also provided for the opening of new careers and opportunities.

TYPES OF CONTROL—OPEN LOOP AND CLOSED LOOP

In computer interfacing applications, the computer and input/output (I/O) boards which make up the DA & C computer-based system can be used in one of two configurations. These two configurations relate to the type of control used in the system; or stated another way, the two possible control systems are related to the inclusion or exclusion of a feedback loop in the control system.

The two types of control are identified as *Closed Loop* or *Open Loop*. An example of an open loop computer system is one that makes use of the time-keeping ability of a computer to turn on the lights in a shopping center parking lot at 8 PM each night. In this application the computer, through the use of its internal DOS clock, monitors the time of day. At 8 PM the computer sends an electrical signal, via a DA & C board, to the relay that controls the parking lot lights. The relay, effectively controlled by the computer, turns on the lights at the programmed time. The relay can be thought of as under the control of the executing software which monitors the computer's internal time-of-day clock. (See Figure 1.3.)

An advantage of this type of computer-based control system is that the computer and its operating software allow for easy revisions to the software program. For example, this flexibility can permit changing the time of day that the lights are turned on and off. The flexibility of the computer also allows for various lights throughout the parking lot to be turned on at different times and off at different times. Perhaps some lights would be turned on at 8 PM while other lights might be turned on at 8:30 PM. Even in a simple application such as controlling parking lot

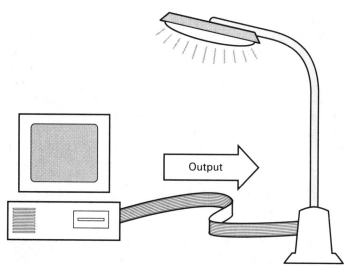

Figure 1.3 Example of Open Loop Control System

lights, an open loop circuit controlled by a personal computer can provide a level of flexibility that is not easily accomplished with a dedicated hardware-based system.

While this open loop application is clever, it does have its limitations. One serious limitation is that the computer never "knows" whether the parking lot is actually dark and, therefore, whether the lights should be turned on. Second, the computer does not have any way of knowing whether the lights really did turn on in response to the electrical signal sent to the relay. Open loop circuits lack *feedback* from the system being controlled.

This lack of feedback to the computer leads system designers to consider "stepping up" to a more sophisticated form of control circuit that will incorporate feedback from the controlled system. A closed loop control circuit offers the same control capabilities as the open loop control system, however, it also allows for feedback, or input, from the "real world." The feedback capability expands the sophistication of the computer system so that the computer is not limited to simply turning on the lights at a predetermined time. It provides for a control system that can incorporate sensors to determine if it is dark enough to turn on the lights. The feedback system can also verify that the lights did turn on as commanded by the computer.

With a modest increase in software programming, along with the use of an interfacing board that has input capability as well as output capability, a closed loop system can be implemented. This type of system can provide intelligent control of shopping center parking lot lights or very sophisticated process control and automation systems. Computer based DA & C systems, using closed loop control, allow system designers to develop control applications that go well beyond the capabilities of traditional dedicated electromechanical clocks and integrated circuit-based systems. (See Figure 1.4.)

Open loop control circuits have many applications and certainly have their place in industry. Where feedback from the controlled circuit or device is not warranted or where cost may be a dominant influence, a simple open loop circuit may be all that is needed for control. On the other hand, situations in which verification of control status is essential and where a higher level of sophistication is warranted, the closed loop control circuit is easily obtainable. The choice of open loop versus closed loop is really one of application.

INTERFACING TO THE REAL WORLD—INPUT SIGNALS

Electrical signals output by sensors fall into one of two categories. Signals produced by sensors (transducers) are either analog or digital. *Analog signals* are continuous and vary in amplitude depending upon the physical phenomenon being measured.

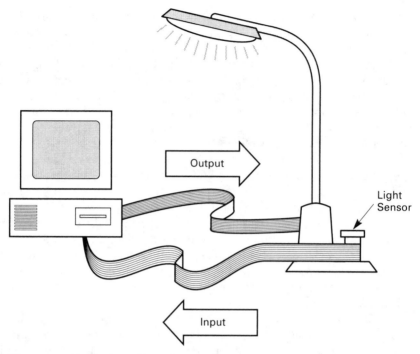

Figure 1.4 Example of Closed Loop Control System

Figure 1.5 illustrates a common analog sensor and its varying output voltage which is influenced by temperature. The transducer shown in Figure 1.5 is a *thermistor*.

Thermistors are resistor-like devices that are commonly used to measure temperature. Thermistors typically have a negative temperature coefficient, which means that as temperature increases, the internal resistance of the thermistor decreases. The accompanying graph in Figure 1.5 illustrates that the voltage produced by the sensor circuit increases as the heat source is moved closer to the sensor. As the temperature of the thermistor increases due to the nearness of the heat the resistance of the negative temperature sensor decreases. This decrease in resistance causes the voltage output from the voltage divider to raise. Movement of the heat source toward the sensor causes an increase in the output voltage, while movement of the flame away from the sensor causes a decrease in the analog voltage produced by the sensor circuit. This analog signal can be interfaced to a computer through an analog-to-digital (A/D) interface board. Through the data acquisition board the computer can monitor the voltage, and therefore, the existence and physical location of the flame.

A variety of analog transducers are used in industry, medicine, and science. Typical of the physical variables capable of being measured by analog transducers

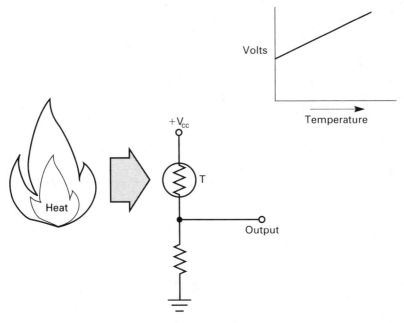

Figure 1.5 Thermistor Circuit with Heat Source—Temperature vs. Voltage

are *temperature*, *force*, *pressure*, *flow*, *displacement*, and *level*. Transducers are manufactured specifically for each physical phenomenon being monitored. A few of these sensors will be more fully described in subsequent chapters.

Digital signals can also be interfaced to DA & C boards. These electrical signals are either ON or OFF in character. They typically exist as a positive value voltage level (perhaps $+5$, $+10$, or $+12$ volts) identified as a logic 1 signal, and typically 0 volts identified as a logic 0. The difference between the two logic levels can be illustrated by the classic light bulb example. This example of the distinction between logic 1 and logic 0 modes uses a light bulb controlled by a toggle switch. When the light bulb is lit the circuit is said to be in the logic 1 (ON) mode and when the light bulb is off in is said to be in the logic 0 (OFF) mode. *Digital transducers* are readily available and can be connected to monitor many different physical situations. One example of a digital sensor application is illustrated in Figure 1.6.

In Figure 1.6 a computer is monitoring one of many doors as part of a burglar alarm in a home or business. Attached to each door is a miniature *reed switch* (transducer) in which the reed switch contacts are held closed by a nearby permanent magnet. If the door is opened, the magnet moves away from the reed switch, and the reed switch springs open. This transition from a closed contact to

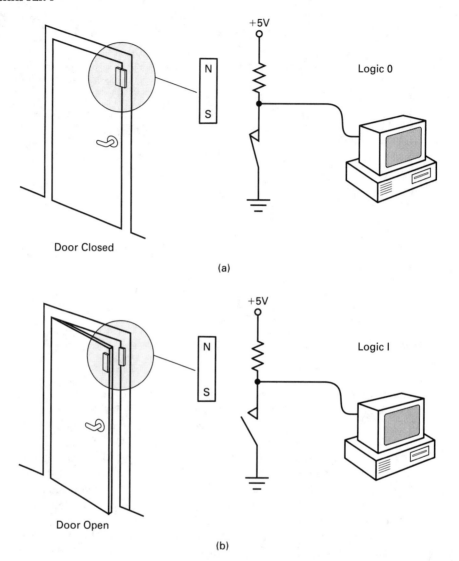

Figure 1.6 Digital Signal created by the Proximity of a Permanent Magnet

an open contact provides for a change in voltage and logic to the DA & C board in the computer. In the drawing of Figure 1.6 the logic level with the door closed is logic 0 (0 volts). When the door opens the logic level changes to logic 1 (+5 volts).

The program execution in the computer in Figure 1.6 is monitoring the status of the logic level coming from each door. The software is written so that the computer will recognize the change in logic level as a door is opened. Depending on how the software was written, the computer could activate an alarm bell, horn, call the police department via a modem and telephone line, or cause another appropriate action such as turn on security lights.

Digital transducers are popular because of their simplicity of operation and generally their direct compatibility with digital computers. If the transition in logic levels is compatible with the logic level used in the digital computer (typically logic 1 = +5 volts and logic 0 = 0 volts), the interfacing of the sensor signal through the DA & C board to the microprocessor in the computer is a simple task. If sensors are used that do not operate on the typical +5 volt level of most computers, then signal conditioning is required to insure that the nonstandard voltage level is translated to the traditional logic level of the computer system. Magnetic Reed Switch sensors of the type shown in Figure 1.6 are used in counting applications, position sensing applications, and a variety of situations where logic 1 or logic 0 status is adequate and the continuous nature of an analog signal is not needed. Digital transducers and signal conditioning will be presented more thoroughly in Chapter 3.

CHAPTER 2

PCs in Data Acquisition & Control Technology

Before the development of the microprocessor, industry relied on human operators for data acquisition and control of processes and equipment. As more sophisticated processes placed greater demands on speed, quality, and volume, the need for reliable precision process monitoring and more precise control became evident. Computer-based process monitoring in use today offers industry a solution to the demand for more sophisticated data acquisition and control systems. Computers can take thousands of readings per second and perform tests and manipulations on the data with much greater precision and reliability than human operators. Additionally, computers never get bored or tired and they rarely "call in sick." Barring an equipment malfunction, computer systems do the same repetitive task with a high degree of accuracy. However, computers did not just happen; they evolved over time.

MICROCOMPUTER HISTORY

The focus of this text is interfacing to the IBM-PC or compatible clone. These computers typically use a microprocessor manufactured by INTEL Corporation. The following chronology describes an approximate history of the evolution of the INTEL family of microprocessors that serve as the CPU in most PCs.

1971. INTEL introduced the *first* generation of microprocessors. These microprocessors, identified as the 4004 (4-bit) and the 8008 (8-bit),* were used initially in calculators and video display terminals. They were designed to replace more complex custom integrated circuits.

1974. The 8008 microprocessor matured into the 8080 microprocessor. The 8080 is characterized as a *second* generation 8-bit processor. Besides electronic logic units such as calculators, the 8080 microprocessor could be used in more complex applications such as word processors and missile control systems.

1976. In response to ZILOG Corporation's entry into the microprocessor field with the Z-80, a competing 8-bit microprocessor, INTEL introduced an enhanced version of the 8080 microprocessor, the 8085. INTEL's ultimate goal was to release a sophisticated processor by the end of the decade. This processor was identified initially as the 8816 but later its identification was changed to the 8800 to compete directly with the Z-80. Development of this new product took longer than expected, so INTEL released its *third* generation processor, the 16-bit 8086.

1979. INTEL released the 8088 microprocessor, a version of the 8086 which has an 8-bit wide external data bus. The 8086 and the 8088 are exactly the same internally and have the same instruction set. However, the 8088 reads and writes data 8-bits at a time. In other words, the "door way" to the chip that the data must pass through is only 8-bits wide. To compensate for this limitation, the processor automatically handles 16-bits and reads and writes as two successive 8-bit operations.

1981. IBM selected the INTEL 8088 microprocessor as the core of its new personal computer (PC). This stroke of good fortune assured INTEL's position as a market leader in the field of microprocessors.

1982. INTEL released the 80186 and 80188 microprocessors. These chips incorporated the 8086 and 8088 CPUs along with some necessary support functions. The capabilities of these newly released microprocessors were the same as the 8086 and 8088 processors. However, fewer chips were needed to accomplish the same task.

INTEL released the 80286 microprocessor. This product was the company's first major step toward microprocessor improvement since the 8086 chip. The main difference between the 8086 and the 80286 is that the newer CPU has better memory management capability.

1985. INTEL released the 80386 microprocessor, a 32-bit processor with enormous speed and memory capability. In the 20 years represented by this chronology, INTEL microprocessors have grown from slow 4-bit processors to lightening fast

---

*When a processor is identified as an 8-bit processor, this means that the standard size of data manipulated by the processor is 8 bits wide. Other processors handle data in 16-bit words, and still other microprocessors handle data in 32-bit words.

32-bit processors. Even with this amazing rate of growth, the technology continues to improve. Along with these well-known processors, INTEL developed a series of *co-processors*, which are devices intended to co-exist in the computer system with the primary processors. The co-processors specialize in fast numeric operations. The 8087 co-processor was designed to work as a companion to the 8086, 8088 microprocessors. The 80287 co-processor was designed to work with the 80286 microprocessor, and the 80387 co-processor was designed to work with the 80386 microprocessor.

1989. INTEL released the 80486 microprocessor. This processor is an improvement over the 80386 microprocessor because it has an internal co-processor. Versions of the 80486 can operate at clock speeds of 66 MHz.

1993. The Pentium processor, the next generation of microprocessors, was introduced by INTEL. This microprocessor is slated to become the "brains" for high-end PC systems.

PERSONAL COMPUTER ARCHITECTURE

Just as microprocessors evolved over time, the personal computer has also evolved through a variety of levels of sophistication. In 1981, IBM introduced the IBM-PC. Though there were personal computers already in existence, the IBM-PC and its bus architecture quickly dominated the market. Computers that used other buses, for example the S-100 bus, were soon at a disadvantage and were dropped from production. At the time of its introduction, the IBM-PC, known as the Model 5150 Personal Computer, was an amazing machine. However, by today's standards it was unsophisticated and limited. The first PCs used cassette tapes for program storage rather than floppy disks and hard disk drives were not available. Sensing consumer demand, IBM quickly included floppy disk drives in updated models of the personal computer. Eventually, even the floppy disk drive was not fast enough nor did it provide enough storage space for users and PC manufacturers added a hard disk drive.

The IBM-XT, built around the 8088 microprocessor, offered hard disk storage along with math co-processor capabilities. The configuration of hard disk storage and floppy disk storage is the standard used for most personal computers today.

The next major improvement in personal computers was the introduction of the IBM-AT. The AT used a 80286 microprocessor which is superior to the 8088 processor that is used in the IBM-PC and the XT. The 80286 has a 16-bit data bus and generally operates at a much higher clock speed compared to its predecessors. In addition to being a faster processor, the AT provides for faster data exchange with disk drives and greater memory capability.

Based on an understanding of how a microprocessor works and how a microprocessor cooperates with other essential parts of a computer, we can see how a personal computer is built around this fantastic integrated circuit. Figure 2.1 shows a block diagram of a typical personal computer.

The processing, or decision-making, ability of the computer is accomplished by components on the main system board. These components include the *microprocessor central processing unit* (CPU), the *random access memory* (RAM), and the *basic input/output system software* often called BIOS. BIOS is typically software, often called *firmware*, inside one or more *read only memory* (ROM) chips.

The microprocessor, in its many variations ranging from the early 8080 to the contemporary 80386, 80486, and Pentium processors, makes an interesting study. These processors are used in a variety of devices other than personal computers. Products ranging from automated factory equipment, to automobiles, and all sorts of consumer products use microprocessors.

Another major component on the computer system board is random access memory (RAM). The RAM operates in cooperation with the CPU. Data written to a memory location remains at that memory location until it is modified or the power to the computer is turned off.

Computers use certain parts of RAM for specific applications. The basic input/output system (BIOS) consists of software which provides instructions needed to initialize the computer when it is first turned on. Several additional routines allow the system hardware to be accessed. Appendix A lists some BIOS functions. These instructions are stored in read only memory (ROM) located on the system board and are addressed as part of the overall memory for the computer. BIOS routines

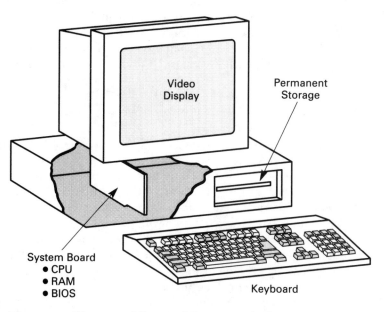

Figure 2.1 Diagram of Personal Computer Architecture

are reached through software interrupts. These BIOS routines can be thought of as the instructions needed to print alphabetic or numeric characters on a CRT screen, output text to a printer, or route data to a communications port. In general, the BIOS accomplishes all of the "low-level" routing work essential to the operation of a computer.

MEMORY AND PORT MAPPING

A memory map is a list identifying locations in a computer's memory that are used for specific functions. Figure 2.2 shows a memory map for the IBM-PC. One of the many functions of the RAM memory is to provide storage for the video display.

Information displayed on the video monitor is stored in the system RAM and is constantly being copied from the RAM to the computer screen. The portion of RAM memory used in this role has an absolute address. In PCs that use monochrome or color graphic adapters, programmers can directly control the text on the screen by storing alphabetic letters in the video RAM. Other system variables along with application program instructions are stored at certain RAM addresses. In Figure 2.2, the screen memory is assigned to RAM addresses ranging from A0000 through C3FFF.

Just as the computer's memory usage can be mapped and addresses assigned to various portions of memory, a computer's input/output (I/O) ports also can be mapped. Figure 2.3 provides a listing of the I/O ports in an IBM-PC. Referring to this figure, note that the first communications port, COM 0, uses ports 3F8 through 3FF hex. The I/O address for a printer is 378 through 37F hex.

These addresses identify the "doorways" through which access can be gained to the communications portion of the computer. As a rule, each device must have its own dedicated port and each port must be used for only one device.

INTERRUPTS

Interrupts provide a method for stopping the normal operation of a computer and causing the computer to redirect its attention to the hardware or software source of the interrupt. For example, if a line printer wants to tell the computer that it is out of paper, the printer causes a hardware interrupt that is communicated back to the computer. A hardware interrupt is an electrical signal to the CPU that informs the computer that one of the hardware devices requests attention.

A software interrupt, on the other hand, is a means for invoking the BIOS and DOS routines that control hardware devices. Programmers may want to access the line printer directly without using a high-level language interface such as LPRINT. Instead, the programmer could invoke a BIOS software interrupt such as 17 hex. The software interrupt would accomplish the same task as the LPRINT command.

Address	Description	
FFFFF 100000	AT Extended Memory (15 Meg)	
FFFFF F0000	ROM	
EFFFF E0000	Open in PC/XT (64 k)	
DFFFF D0000	LIM Expanded Memory	
CFC00 CF800 CF400 CF000 CEC00 CE400 CE000 CDC00 CD800 CD400 CD000	User Area	
CCFFF C8000	Fixed Disk XT (20 k)	
C7FFF C4000	ROM Exapansion	
C3FFF C0000	Open (16 k)	
BFFFF B0000	CGA Screen Buffer	EGA Buffer
AFFFF A0000	Open (64 k)	
9FFFF 80000	RAM Expansion (128 k)	
7FFFF 00400	RAM Expansion (512 k) BIOS	
003FF 00000	Interrupt Vectors	

Figure 2.2 IBM-PC Memory Map

Hex range	Usage	
000-00F	DMA chip 8237A-5	
020-021	Interrupt 8259A	
040-043	Timer 8253-5	
060-063	PPI 8255A-5	Assigned to
080-083	DMA page registers	system board
0Ax	NMI mask register	components
0Cx	Reserved	
0Ex	Reserved	
100-1FF	Not usable	
200-20F	Game control	
210-217	Expansion unit	
220-24F	Reserved	
27S-27F	Reserved	
2F0-2F7	Reserved	
2F8-2FF	Asynchronous communications (2)	
300-31F	Prototype card	
320-32F	Fixed disk	
378-37F	Printer	Assigned to
380-38C	SDLC communications	feature card
380-389	Binary synchronous communications (2)	ports
3A0-3A9	Binary synchronous communications (1)	
3B0-3BF	IBM monochrome display/printer	
3C0-3CF	Reserved	
3D0-3DF	Color/graphics	
3E0-3F7	Reserved	
3F0-3F7	Diskette	
3F8-3FF	Asynchronous communications (1)	

Figure 2.3 I/O Port Map of an IBM-PC *(From Tompkins & Webster, Interfacing Sensors to the IBM PC, Prentice-Hall, 1988)*

CONTROLLING THE HARDWARE

Hardware within the computer can be controlled by using different levels of commands. The lowest level of control is *direct hardware manipulation*. The next higher level of control is through *BIOS calls* or *software interrupts*. The third level of control is through the *disk operating system* (DOS) using software interrupts. The fourth and highest level of control is through a *high-level language* such as C, BASIC, Pascal, and the like. Instructions in each of these levels allows control of the computer hardware including the CPU or microprocessor. Each of these levels of control offers advantages and disadvantages. Controlling the computer through a high-level language will result in slower response; however, there is a greater chance that this instruction will operate correctly on a variety of IBM-PC clones.

For the system hardware to be "compatible" with an IBM-PC, all control and data registers for each hardware device must have the same addresses as those used by IBM. Additionally, each device must be designed to operate in exactly the same way. This places several restrictions on hardware compatibility. Keep in mind that all clones are not 100 percent compatible!

THE EXPANSION BUS

In Figure 2.1, the video monitor, the keyboard, and the disk drive are all external hardware devices. IBM designed the personal computer so that the user could have some freedom in the selection of video monitors, disk drives, and other optional equipment. Manufacturers made this possible by including an expansion interface bus on the system board. Figure 2.4 shows an IBM-PC expansion connector.

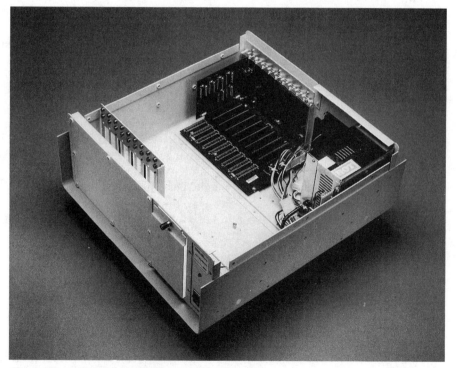

Figure 2.4 IBM-PC Expansion Bus Connector

The expansion bus is a series of edge connectors, one for each expansion slot. These connectors include all of the signals, data, address, and control lines necessary for the operation of the expansion boards that plug into the slot. The original IBM-PC and PC-XT bus had 62 contacts. When the IBM-AT computer was developed, an additional group of contacts were added to the expansion connection. The new section has 36 contacts which provided the AT with expanded memory and a 16-bit data bus. The right-hand portion of Figure 2.4 shows the AT bus connector including the 36 additional bus contacts that are unique to the AT bus.

Figure 2.5 illustrates cards plugged into the expansion slots inside a typical PC. Figure 2.5a shows removal of the cover, Figure 2.5b illustrates cards installed, and Figure 2.5c shows a view of the *mother board*, often called a *backplane*. Notice that in Figure 2.5c the manufacturer has provided two PC expansion slots, two AT expansion slots, and four slots designed for local bus cards.

EXTERNAL BUS DATA ACQUISITION

Technology continues to become more complex with greater demands placed on the equipment, programmers, and systems engineers. Yet computers provide fast and reliable process monitoring and manufacturers such as INTEL are producing more powerful processors to meet this demand.

The method by which computers are integrated into the individual control system varies. Computer-based data acquisition and control can be divided into two classes: *external bus data interfacing* and *internal bus data interfacing*. Each of these classes has its advantages and disadvantages.

External bus data interfacing connects the computer with interfacing circuitry, such as A/D converters and other signal conditioning circuitry, through a standard communications channel such as RS-232, RS-422, or IEEE-488. The external data interfacing circuitry is located external to the computer, often at a remote location. All of the circuitry needed to perform the data acquisition is external to the computer. Only the data is sent through the communications channel to the host personal computer.

External bus data acquisition has several advantages. The following list summarizes the advantages.

1. Since the communication channels (RS-232, RS-422) used between the data acquisition hardware and the host computer are standardized, the remote equipment can be connected to almost any host computer.

2. Because the data collection equipment is self-contained, it can be placed near the source of the signals in remote locations.

3. Some data collection tasks can be programmed into the remote unit, freeing the host computer for other tasks.

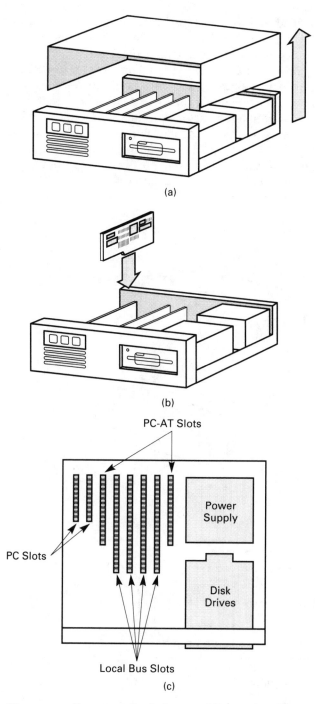

(a)

(b)

PC-AT Slots

Power
Supply

PC Slots

Disk
Drives

Local Bus Slots

(c)

Figure 2.5 Computer Backplane and Expansion Slots

The reader is reminded that when external data acquisition is used, only the data is sent to the host computer, not the actual field electrical signals.

Figure 2.6 shows an external bus data acquisition system in block diagram form. Notice that the data acquisition system is external to the PC. Proximity to the source of the field signals can greatly reduce electrical noise that can cause data errors. A close look at Figure 2.6 reveals that the external data acquisition system contains a microcomputer, actually a microprocessor or microcontroller. Inclusion of a microprocessor or microcontroller in the data acquisition hardware allows programming of the external data acquisition device. Programming of the remote data acquisition system relieves the host computer from most of the data handling and allows the host computer to do other tasks such as data analysis, graphical presentation, or another function while the external unit is gathering new data.

External bus systems do have some disadvantages. The following list identifies some of the disadvantages.

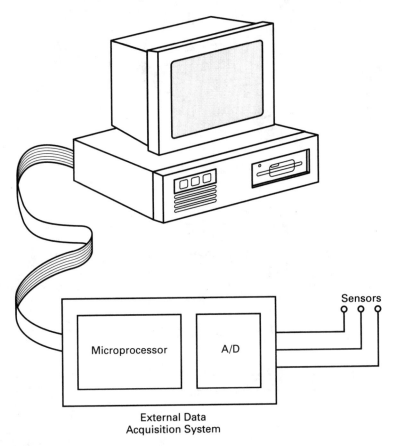

Figure 2.6 External Bus Data Acquisition

1. They tend to be more expensive than internal systems because of added hardware (enclosures, power supplies, etc.)

2. They are slower than internal systems. While data can be collected at a fast rate, transmission of the digitized information to the host is limited by transmission speeds. Even at relatively high transmission speeds (9600 baud) the computer has to "wait" for the data from the external data acquisition system.

INTERNAL BUS DATA ACQUISITION

Internal bus systems are characterized as boards that plug into sockets in the PC's expansion bus. Figure 2.7 shows an internal bus data acquisition board installed in the expansion bus of a personal computer.

The board contains only the data acquisition circuitry and does not have a separate microprocessor or microcontroller installed. Signals are brought from the sensor or other transducer device to the back of the PC where they are connected to the data acquisition board. All tasks related to the control of the I/O device and all data acquisition is the responsibility of the host microprocessor.

Internal bus systems have some distinct advantages. These advantages are:

1. *Lower costs*. The host computer provides power, computational ability, and data storage.

2. *Smaller size*.

3. *High speed*. At 9600 baud, the external system can read approximately 20 analog channels per second, while in the same amount of time the internal bus system can read more than 100,000 channels per second.

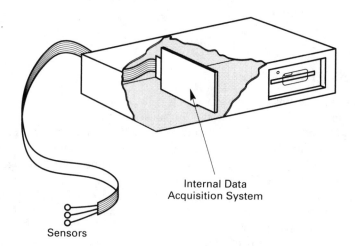

Internal Data
Acquisition System

Sensors

Figure 2.7 Internal Bus Data Acquisition Board

The remainder of this text will be devoted to internal bus data acquisition devices. Internal bus I/O boards are maufactured by a number of companies including Keithley MetraByte Corporation, 400 Myles Standish Boulevard, Taunton, MA 02780. Before discussing the interfacing boards themselves, a firm understanding is needed of how an electrical signal created by a sensor is fed to the data acquisition board in the computer.

GETTING DATA INTO THE PERSONAL COMPUTER

Chapter 1 introduced the idea that electrical signals originating from external sensors may be input to a personal computer via the DA & C. The chapter also defined these signals as either digital or analog in character. Analog signals will take the form of varying voltage or current levels while the digital signals will conform to the typical ON or OFF (logic 1 or logic 0) status common to any digital logic signal. While either type of signal can be connected to a PC, it is necessary to have a DA & C board that is compatible with the signal or signals being applied. Therefore, an analog input signal must be connected to an analog data acquisition and control board and a digital input signal must be connected to a digital data acquisition and control board. Some DA & C boards can receive both analog and digital signals. Analog voltages or currents are applied to the analog portion of a multifunction board and digital signals must be applied to the digital portion of the board.

Both the analog signal and the digital signal may, and in some cases must, conform to an industry standard protocol. Protocol means that the electronics industry, typically the *Institute of Electrical Electronics Engineers* (IEEE) or other group such as the *Electronics Industries Association* (EIA), has standardized voltage or current levels and connecting pin configurations that must be used for system and component compatibility.

ANALOG AND DIGITAL STANDARDS

Analog transducers produce either a varying voltage or current in response to the physical phenomenon that is influencing the sensor. To assure a uniform range of varying voltages and/or currents that are produced by sensors, the electronics industry has established common voltage ranges or current ranges that allow the "standardization" of most analog signals. Common analog voltage ranges, often called *spans*, are 0 to +5 volts, 0 to +10 volts, and −2.5 volts to +2.5 volts. A particularly common analog current range is the 4 to 20 mA current loop standard. The *current loop standard* defines an analog signal that is transmitted from the transmitter to the receiving device, computer, recorder, and so on. The standard definition of a *full-scale span* is for currents to vary between the limits of 4 mA to 20 mA (16 mA span).

Digital standards also exist and are generally much simpler than the analog standards. Digital standards are based on the 0 volt and + 5 volt logic levels common to the TTL logic family. Occasional variations exist; however, the 5 volt TTL standard is so common that it is almost "carved in stone."

Chapter 3 will include a more detailed look at both analog and digital signals and will provide more information regarding these standards.

Real World Sensors and Input Signal Conditioning

Sensors are an essential part of data acquisition and control systems. *Sensors*, often called *transducers*, are devices that when acted upon by physical variables such as force, pressure, temperature, flow, and level, produce electrical signals. The signals produced by the sensor are proportional to the amount of pressure, flow, or temperature being monitored. This electrical signal is connected to the computer or other electronic system that is serving as the basis of the data acquisition and control system. (See Figure 3.1.)

Sensors can be categorized several ways. One method of categorizing a sensor is according to the physical phenomenon that the sensor is measuring. This method allows the grouping of all forms of pressure sensors into a single category, and the grouping of temperature sensors into another category. Another way of categorizing sensors is whether the sensor is either a passive or active device. A *passive sensor* is a device in which the electrical energy created by the sensor is generated by the physical phenomenon being measured. Typical of this type of sensor is a thermocouple. *Thermocouples* are used to measure temperature. Another passive sensor is the light sensitive *photovoltaic photo-cell*. Both of these devices are typical of passive devices in which the electrical signal is "created by" temperature, light, or other physical variable.

Active sensors are devices that require an external "excitation" voltage applied to the sensor device. Active devices usually, though not always, have three wiring connections to the sensor. The physical phenomenon being measured varies (modulates)

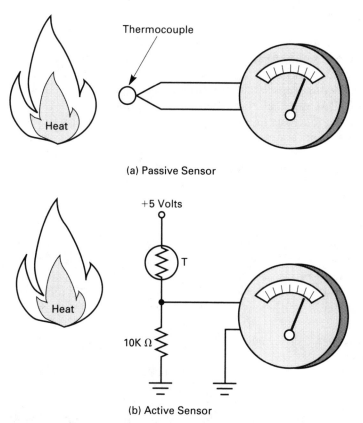

Figure 3.1 Passive and Active Transducers

the supply voltage. The value of the changing electrical signal output from the sensor is related to the status of the physical variable being measured. As the pressure, temperature, or other physical property changes the electrical signal varies proportionally.

The types of electrical signals created by transducers must be defined. These signals can be analog or digital. While all these signals are assumed to be changing in concert with the varying physical phenomenon being measured, the analog signal is the only signal that accurately reflects the degree of change or the status of the measured phenomena.

Generally, most sensors create an analog signal at the point of measurement. In a typical industrial application, the point of measurement will be in the factory, some distance away from the computer. The analog output from the sensor will be converted into a digital format either at the point of measurement, by an external DA & C system, or in the host computer by using an internal DA & C system. In either case, the analog signal must be converted to a digital format so that the microprocessor can interpret the measured data. This conversion is accomplished by a *digital-to-analog (D/A) converter*.

TEMPERATURE SENSORS

Temperature is one of the most common physical phenomena measured. A variety of sensors are designed to specifically measure temperature and provide an analog output. Though most *temperature sensors* produce an analog output there is one type of temperature sensor that produces a digital output. This sensor, perhaps the simplest of all sensors, uses the thermal expansion coefficients of two different metals to provide an opening or closing "snap action" when the temperature of the sensor rises above a predefined value. *Snap action bimetallic switches* are practical in situations where continuous analog readings are not necessary and where the temperature change occurs slowly. (See Figure 3.2.)

Thermocouple

One of the most common analog temperature sensors is the *thermocouple*. This sensor is rugged, reasonably priced, and can monitor temperatures from extreme cold, −200 degrees Celsius, to 3000 degrees Celsius. A thermocouple, shown in Figure 3.3, is a two-wire device made up of two dissimilar metals with one end welded together. When heat is applied to the welded junction, a voltage is generated by the dissimilar metals. The thermocouple is a positive coefficient device. As the temperature rises, the generated voltage rises, and as the temperature falls the voltage falls.

Many types of thermocouples are available. Each type is manufactured from different metals and is intended for different temperature ranges and applications. For convenience and standardization, alphabetic labels have been given to each type of thermocouple. Typical of the labels given to the different types of thermocouples are the designations: Type J, Type K, and Type T. Each thermocouple type has a slightly different electrical and temperature characteristic and, therefore, a different use. When selecting a thermocouple, the following general characteristics should be considered.

Type J	Low Cost	Should not be used above 760 degrees Celsius.
Type K	Moderate Cost	High temperature range. Can be used up to 1000 degrees Celsius.
Type T	Moderate Cost	Useful at low to moderate temperatures ranging from −160 to 400 degrees Celsius.

Unfortunately the voltage output from a thermocouple cannot be measured directly using a voltmeter. The voltage level from the thermocouple is very small. The small signal level along with other factors require that compensation junctions, or ice baths, be provided. These factors make the direct measurement of the output from a thermocouple impractical. The thermocouple application in Figure 3.1 should be interpreted as a conceptual picture, not an actual schematic.

Low Level Hermetic Thermostats

Series 3106

The Series 3106 thermostat is designed for use with logic level circuits. Model 3106U is similar to the 3106 but manufactured and tested for applications that require a UL rating.

The 3106 is a single-pole, single-throw switch activated by a snap-action bimetal disc. Temperature calibrations are pre-set at the factory, and each unit is 100% thermally and mechanically inspected. It is available to open or close on temperature rise. The case is laser welded to form a hermetically sealed steel housing, with a glass-to-metal seal at the terminal junction.

To insure that a safe combination of thermostat and application is achieved, the purchaser must determine product suitability for their individual requirements.

Key Features and Benefits

- Hermetically Sealed
- WE-1 Gold Alloy Cross Point Contacts
- Narrow or Wide Differentials
- Environmental Exposure −80° to 500°F (−62° to 260°C)
- UL Recognized
- Single-Pole, Single-Throw (SPST)
- Pre-set and Tamperproof
- Variety of Mounting Brackets and Terminals Available

Typical Applications

- Logic Level or Dry Circuit Applications

Internal Configuration

```
              Terminals
   Glass Header

Gold Alloy                          Housing
Contacts
                                    Contact Arm

                                    Ceramic Insulator

                                    Laser Weld
     Cap            Bimetal Disc
        Ceramic Transfer Pin
```

Figure 3.2(a) Bimetal Switch and Sample Digital Application

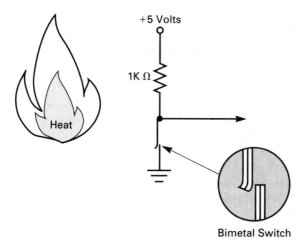

Figure 3.2(b) Bimetal Switch and Sample
Digital Application

Fortunately, the awkward situation of requiring compensation junctions and conditioning circuits is eliminated by using DA & C boards that are specifically designed for use with thermocouples. These boards provide the compensation needed to allow the direct connection between the two-wire thermocouple and a computer. Using a DA & C board that has been specifically designed to accommodate thermocouples makes the use of a thermocouple as simple as connecting two wires between the sensor and the computer.

Thermistor

Another temperature sensor used frequently is the thermistor. This device is a temperature sensitive resistor which changes resistance proportionally to temperature. Most thermistors have a negative temperature coefficient though positive coefficient thermistors are available. Thermistors are made from semiconductor material, and they have a high level of sensitivity. The sensitivity characteristic allows the thermistor to respond quickly to temperature changes. (See Figure 3.4.)

The disadvantage of the thermistor is its nonlinearity. This characteristic can be corrected using mathematical calculations to correct the nonlinear response. Thermistors are active devices and require an excitation voltage. Typical of a thermistor sensor circuit is the circuit shown in Figure 3.1(b). This circuit provides an output voltage proportional to the ambient temperature surrounding the thermistor. As the temperature increases, the resistance of the thermistor will decrease. A resistance decrease in the thermistor portion of the voltage divider will cause the voltage applied to the voltmeter to increase. Thus, as the temperature increases the voltmeter needle will move toward the right side of the meter. As the temperature decreases the needle will move toward the left side of the meter.

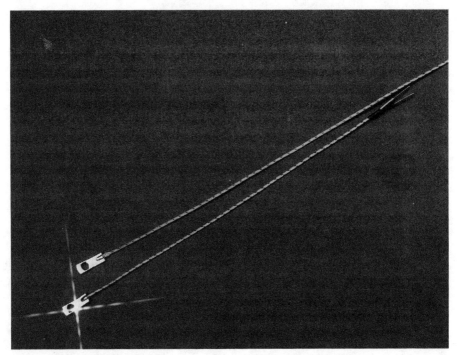

Figure 3.3 Thermocouple

Solid State Temperature Sensors

The newest form of temperature sensor is the monolithic linear temperature sensor. This sensor is derived from silicon integrated circuit technology and is often called an integrated-circuit sensor. This device is available in a TO-92 transistor style, the SO-8 plastic 8-pin mini-DIP integrated circuit, or the TO-46 metal can package. The most common of the solid state temperature sensors is the LM34 produced by National Semiconductor Corporation, 2900 Semiconductor Drive, Santa Clara, CA 95052. The LM34 is available calibrated for Centigrade or Fahrenheit temperatures.

The voltage output from this device is accurate for a wide range of temperatures and the output voltage possesses a high degree of linearity. The usable temperature range for a typical LM34 sensor is − 50 to + 300 degrees Fahrenheit. Figure 3.5 illustrates a typical solid state temperature sensor along with the power supply, ground, and signal output connections. Solid state temperature sensors like the LM34 are particularly popular because they are accurate, linear, inexpensive, and easy to use. Their limitation is that they can monitor only moderate temperatures; and like most semiconductor devices, they are not "forgiving" if connected in reverse polarity or to the wrong voltage. Appendix B provides additional information on the LM34 temperature sensor.

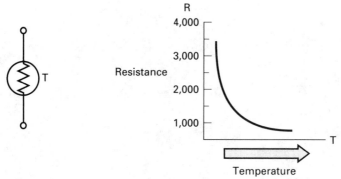

Figure 3.4 Thermistor Symbol and Characteristic Curve

LIGHT SENSORS

A variety of optical sensors are available to the technician and engineer. Typically, an optical sensing system involves both a sensor and a light source. The phenomena being measured, such as smoke, moving mechanical parts, an opaque liquid, or boxes on a conveyor belt, diminish or block the amount of light entering the sensor from the light source. The sensor converts the varying intensity of the light to a corresponding electrical signal that is fed to the DA & C board.

Light sensors are divided in to two major catagories. One of the categories is identified as photoconductive and the second is photovoltaic. Photoconductive sensors are the most widely used form of light sensor. These sensors are often called light-activated resistors or light-dependent resistors (LARs or LDRs).

As the amount of light enters the LDR sensor, its resistance changes. Typically, these devices exhibit a negative resistance coefficient. Figure 3.6 shows an LDR circuit having a negative coefficient. As the amount of light entering the sensor decreases, the resistance of the sensor increases. Given the configuration shown in Figure 3.6, the analog voltage output from the circuit will decrease as the level of light entering the LDR decreases.

The other type of light sensitive transducer is the photovoltaic sensor. Photovoltaic sensors are known for their ability to convert sunlight into electrical energy. These devices produce a voltage proportional to the amount of light entering the sensor. This form of sensor falls into the passive category as the voltaic device does not require an excitation voltage. Photovoltaic panels are used in applications as varied as powering satellites in space to operating portable calculators.

Phototransistors

A phototransistor also makes an excellent light sensor. Figure 3.7 illustrates a typical phototransistor and an application using this transistor. In this example circuit the phototransistor provides a digital output depending upon the on or off status of the light entering the phototransistor. Light entering a transparent opening in the transistor's housing causes the transistor to be forward biased. Removal or blockage

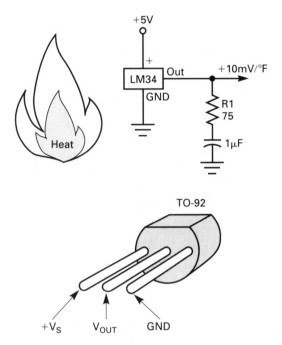

Figure 3.5 Precision Solid State Temperature
Sensor—LM34

of the light causes the transistor to be in the cut-off mode. The change in conductivity between saturation and cut-off depends upon the existence of light hitting the phototransistor. The circuit shown in Figure 3.7 operates in the following manner. Light hitting the transistor causes the device to be in saturation mode. Saturation causes the input to the inverting schmitt trigger to be approximately +5 volts or logic 1. If the light is blocked from hitting the phototransistor, the transistor changes to cut-off mode and the voltage level applied to the inverting schmitt trigger will decrease to approximately 0 volts, logic 0. This type of sensing circuit and interface is popular when physical objects moving along a conveyor belt are counted, or otherwise sensed. Some phototransistors are manufactured that respond to infrared light. Others are produced that respond to visible light. In either case, the transistor offers high sensitivity to changes in the level of light entering the sensor.

PRESSURE TRANSDUCERS

Pressure sensors use the idea of a strain gage to convert pressure into a varying electrical signal. Contemporary analog pressure transducers use an integrated circuit connected to a pressure sensitive diaphragm. The sensor circuit produces voltage

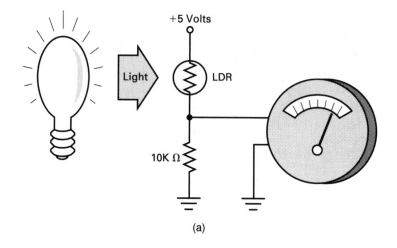

(a)

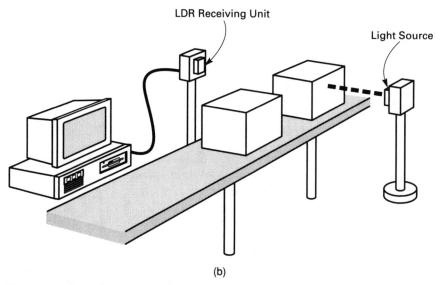

(b)

Figure 3.6 **Light Dependent Resistor—LDR**

or current output proportional to the pressure of liquid, gas, or other medium applied to the pressure input port on the sensor.

Integrated circuit sensors are typically available in two types or styles. One type of transducer produces an output voltage that ranges from 0 volts to $+25$ mV depending upon the pressure applied to the sensor. This form of sensor produces a small signal and needs amplification before being applied to a digital-to-analog converter in a DA & C board.

Another type of pressure transducer, often called a conditioned transducer, produces a signal that ranges from 0 to $+6$ volts depending upon the pressure applied

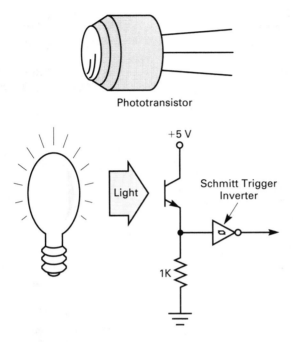

Figure 3.7 Phototransistor Application

to the sensor. Output voltage from the conditioned sensor has been amplified internally by circuitry inside the housing of the sensor. An example of a conditioned pressure sensor capable of measuring 1 psi (27.68 inches of water) is shown in Figure 3.8. The signal output from this sensor would be approximately +6 volts when the water tank is full and approximately zero (0) volts when the tank is empty.

The SenSym series 142SC used in Figure 3.8 is also available in higher pressure ratings up to 150 psi. Appendix C provides additional information on integrated circuit pressure sensors. Integrated circuit pressure sensors make excellent liquid level monitoring transducers due to their linearity, low cost, and ease of installation.

CURRENT LOOP

When analog sensors are located a considerable distance from the computer or controller, it is common practice to transmit the analog signal as current rather than as a voltage signal. The typical standard associated with a current loop circuit is the 4 to 20 mA circuit, though other standards such as 10 to 50 mA do exist. In the 4 to 20 mA configuration, the current ranges from a minimum of 4 mA to a maximum of 20 mA depending upon the output from the sensor. The conversion of a sensor's signal to the varying current is generally provided by a special module identified as a transmitter.

142SC Series
0–1psi to 0–150psi
Signal Conditioned
Pressure Transducers

FEATURES

■ **Improved Performance Replacement for Honeywell/Microswitch 140PC Series**

■ **High Level Voltage Output**

■ **Field Interchangeable**

■ **Calibrated and Temperature Compensated**

APPLICATIONS

■ **Medical Equipment**

■ **Barometry**

■ **Computer Peripherals**

■ **HVAC**

GENERAL DESCRIPTION

The 142SC series transducers provide a 1–6V output which is directly proportional to applied pressure. This series consists of eight (8) devices for monitoring differential, gage, or absolute pressures from 0–1 to 0–150psi. These products feature a high level voltage output, complete calibration and temperature compensation.

Based on Sensym's precision SX series sensors, the 142SC series is an improved performance, direct replacement for the Honeywell/Microswitch 142PC series with equivalent pinout and package mounting dimensions.

This allows direct replacement in existing PC board layouts for the Microswitch parts. Sensyms 142SC devices offer the added advantage of tighter tolerances which give greater accuracy and field interchangeability.

These products are designed to be used with non-corrosive, non-ionic gases and liquids. For more demanding or corrosive media applications, Sensym's ST2000 stainless steel isolated family should be used.

Figure 3.8(a) Pressure Sensor and Water Tank Application *(Courtesy SenSym, Inc.)*

FUNCTIONAL SPECIFICATIONS

142SC Series

Maximum Ratings

Supply Voltage	$+7V_{DC}$ to $16V_{DC}$
Output Current	
Source	10mA
Sink	5mA
Temperature Ranges	
Compensated	$-18°C$ to $+63°C$
Operating	$-40°C$ to $+85°C$
Storage	$-55°C$ to $+125°C$

Reference Conditions

Supply Voltage	$8.0 \pm 0.01V_{DC}$
Reference Temperature	25°C
Common-mode Pressure	0psig

INDIVIDUAL OPERATING CHARACTERISTICS

Sensym Part #	Operating Pressure Range	Proof Pressure	Sensitivity
142SC01D	0–1psid (g)	20psig	5V/psi
142SC05D	0–5psid (g)	20psig	1V/psi
142SC15A	0–15psia	45psia	333mV/psi
142SC15D	0–15psid (g)	45psig	333mV/psi
142SC30A	0–30psia	60psia	167mV/psi
142SC30D	0–30psid (g)	60psid	167mV/psi
142SC100D	0–100psid (g)	200psid	50mV/psi
142SC150D	0–150psid (g)	200psid	33mV/psi

PERFORMANCE SPECIFICATIONS (For All Devices) (Note 1)

Parameter	Min.	Typ.	Max.	Unit
Offset Calibration (Note 2)	0.95	1.0	1.05	V
Output at Full Pressure	5.90	6.0	6.10	V
Full-scale Span (Note 3)	4.95	5.0	5.05	V
Linearity ($P_2 > P_1$)	—	0.5	1.5	%FSO
($P_2 < P_1$) (Note 4)	—	0.2	0.75	%FSO
Temperature Shift ($-18°C$ to $+63°C$) (Note 5)	—	0.5	1.0	%FSO
Repeatability and Hysteresis	—	0.2	—	%FSO
Response Time	—	0.1	1.0	ms

Figure 3.8(b) Pressure Sensor and Water Tank Application *(Courtesy SenSym, Inc.)*

Figure 3.9 illustrates a water level (pressure) transducer in a 4 to 20 mA application. The pressure sensor is connected to the transmitter which converts the sensor signal into a signal appropriate for the receiving device. In this example, the transmitter produces a current output ranging from 4 to 20 mA depending upon the level of the water in the tank.

In Figure 3.9, if the water level is at the top of the tank, the current output from the transmitter will be 20 mA. As the water is drained from the tank the current flow from the transmitter to a DA & C board in the computer will decline

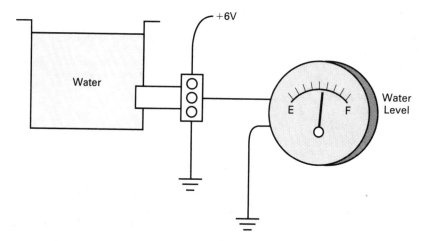

Figure 3.8(c) Pressure Sensor and Water Tank Application *(Courtesy SenSym, Inc.)*

as the level of the water is lowered. When the tank is empty, the current output from the transducer will hold steady at 4 mA. Thus, the range of current flow, 4 mA when empty to 20 mA when full, provides a DA & C board or other instrument with an analog indication of the depth of water in the tank.

Sensors and transmitters which provide variable current in the 4 to 20 mA range must be connected to devices which are capable of receiving the 4 to 20 mA current. A wide range of DA & C boards are available that accept this form of analog input. If 4 to 20 mA signal is not compatible with the DA & C interface in a computer the current signal can easily be converted to a voltage signal for connection to voltage instruments such as voltmeters and voltage input terminals on DA & C boards.

If the designer prefers a more traditional 0 to +5 volt signal, a precision 250 ohm resistor can be placed in the current loop. This resistor will convert the varying 4 to 20mA of current into a +1 to +5 volt signal. The current loop form of analog signal transmission from sensor to receiving device is popular because the current loop provides a high degree of noise immunity as well as a lack of loading difficulties caused by more than one receiving device in the loop. Therefore, a 4 to 20 ma loop can be connected to a DA & C board and, simultaneously, can be connected to a strip chart recorder, meter, or other instrumentation device.

Current loop analog transmission is unaffected by the distance between the transducer/transmitter and the receiving devices. This characteristic makes the current loop method of transmitting analog information very popular in factories, electrical power generating plants, and other locations where analog signals need to be transmitted long distances.

Another desirable characteristic of the 4 to 20 mA standard is that a minimum current flow of 4 mA is always in the circuit. This characteristic can be used as a

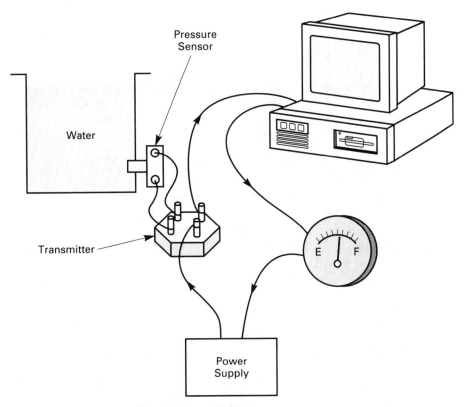

Figure 3.9 4 to 20 mA Transmitter Application

check of circuit integrity. In the water tank example shown in Figure 3.9, the 4 mA minimum amount of current provides for a distinction between zero water in a water tank and zero current. A software program can be written that identifies zero current in the loop as an indication that the cabling from the sensor to the computer is broken or otherwise damaged.

ANALOG VOLTAGE STANDARDS

For those applications where it is desirable to interface a varying analog voltage signal rather than varying current levels, standard voltage ranges do exist. One standard analog signal level is the 0 to +10 volt range span. The standardization of 10 volts as the maximum signal level and 0 volts as the minimum voltage level is arbitrary; other levels do exist. Typical standard analog voltage ranges are:

0 to +5 volts
+/− 5 volts

+/− 2.5 volts

+/− 500 mV

+/− 50 mV

The variation in voltage levels produced by the transducer circuit is dependent upon the supply voltage used in the circuit. Figure 3.10 illustrates a circuit that uses a light dependent resistor (LDR) as an analog sensor.

In this example, the value of the output voltage when the LDR is completely darkened (covered) never exceeds the supply voltage of +5 volts. Shining a bright light into the LDR causes the resistance to decrease to a low value. This decrease in resistance causes the output voltage to fall to a low value. This type of circuit may be used effectively for monitoring levels of light and can form the basis of a dusk-to-dawn light control system.

The use of varying analog voltage levels as inputs to DA & C boards and other forms of instrumentation is acceptable if the transducer and receiving device

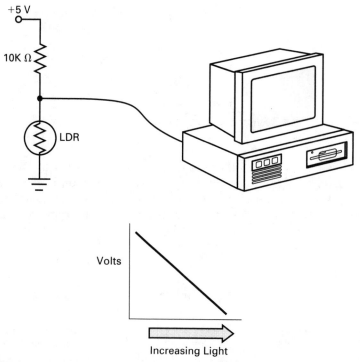

Figure 3.10 LDR Analog Sensor Application

are close to each other. Analog voltage signals are susceptible to electrical noise and loading of the transmission line between the sensor and monitoring device. If the distance between the sensor and the computer or other device is a short, direct connection, the voltage signal may be used with success.

DIGITAL STANDARDS (TTL)

The easiest digital signal to interface to a personal computer via a DA & C board is a TTL (transistor transistor logic) level signal. Industry standards describe this signal according to the following voltage and logic levels:

logic 1   =   +5 volts

logic 0   =   0 volts

The circuit shown in Figure 3.11 is typical of the type of circuit that may be used to produce digital TTL signals. The TTL signal created by this circuit is a logic 1 when the toggle switch is open and logic 0 when the switch is closed. The toggle switch can be opened and closed by a variety of mechanical devices or forces. Examples of devices that open and close the switch are a door opening or closing, the level of water in a tank tripping a mechanical float switch, or temperature in an oven reaching a predetermined temperature level.

The TTL level digital signal is compatible with most digital DA & C boards; however, a limitation is that the signal can be transmitted only a short distance between the sensor and the computer. Transmission distances of a digital signal vary depending upon several factors. Under any circumstance, transmissions are limited to roughly 10 feet. If the signal source and DA & C board are more than a few feet apart, the transmitted voltage signal is likely to be attenuated during transmission. Attenuation will cause a lack of compatibility between the sensor and the signal voltage standards required by the DA & C board. Digital signals of this type are used to express the ON or OFF status of a transducer rather than the relative magnitude of the force affecting a sensor. Magnitude of a force is indicative of an analog signal.

A digital signal connected or applied to a DA & C board may be a single bit as shown in Figure 3.11a or the digital signal may be multiple bits connected in parallel as shown in Figure 3.11b. Parallel digital inputs to the computer via a DA & C board are frequently 8-bits in parallel. The 8-bit wide digital data input to the computer is called a byte. Some DA & C boards provide for multiple bytes of digital input or output. One common digital I/O board, the PIO-12 produced by MetraByte Corporation has 24 bits of TTL I/O that is divided into three 8-bit bytes.

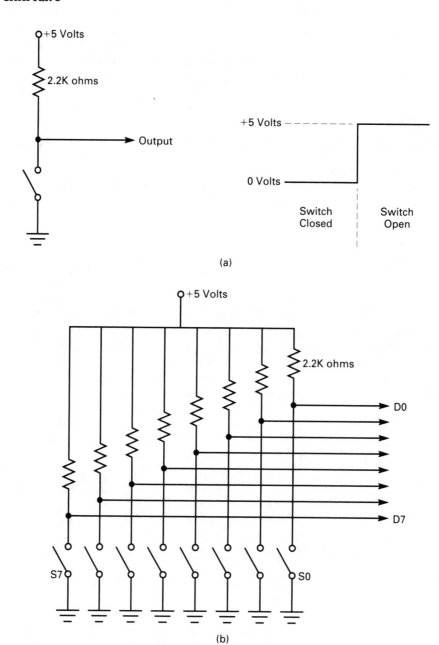

(a)

(b)

Figure 3.11 Digital TTL Level Signals

SERIAL DIGITAL SIGNALS (RS-232, RS-422)

Serial data transmission is a form of digital communication that is used for communicating between sensors and computers, between computers, or between computers and peripheral devices such as printers, plotters, or other output devices. This form of communication between devices uses the same concepts of digital data transfer that were described in the TTL section of this chapter. While the digital concepts (logic 1 and logic 0) are consistent, there are a few differences which make serial digital transmission unique. One major difference between serial and TTL transmission is that serial communication transmits digital signals bit-after-bit sequentially rather than in parallel. For example, transmitting 8-bits of data **serially** requires two wires (one going and one common ground) instead of eight signal wires in parallel along with a common ground. Having fewer wires connecting two devices is an advantage. However, the serial transmission of data takes more time when compared to digital TTL transmission. The longer transmission time occurs because each individual bit takes its turn on the wire. By comparison, in the TTL mode of transmission all 8 bits are transmitted at once.

A major advantage of serial transmission is its ability to transmit signals over long distances. This characteristic is provided through the use of, and adherence to, industry-wide serial communications standards. The most common standard for digital serial transmission is identified by the designation RS-232. The RS-232 standard was introduced in 1962 by the Electronics Industries Association (EIA) and has been updated several times since then by the EIA.

The RS-232 standard defines the voltages that must be used for the transmitted digital information along with the function of connector pins. Figure 3.12 illustrates the voltage ranges and corresponding logic levels for the RS-232 standard.

This standard defines a voltage between +3 and +15 volts as a logic 0 and voltages between −3 and −15 volts as logic 1. Naming a negative voltage a logic

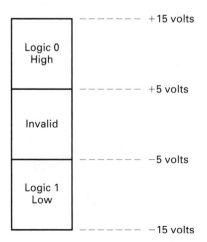

Figure 3.12 RS-232 Voltage Levels

1 or "true" state may seem a bit unusual, but this is the standard! Keep in mind that the designation of 0 volts as a logic 0, and +5 volts as a logic 1 is a convention and is varied by designers and manufacturers of equipment. The important point to remember is that no matter the voltage level, a logic 1 is the "true" state and Logic 0 is "false." Within the RS-232 standard the true state is a negative voltage and the false state is a positive voltage!

Although the use of +/− voltages may seem a bit unusual in adigital application, this form of digital transmission is advantageous because of its high level of noise immunity. The noise threshold for the RS-232 signal is clearly +3 volts and −3 volts for a span of 6 volts. By way of comparison, a check of TTL noise immunity threshold values reveals that TTL logic has a noise immunity of approximately 1.6 volts. Compared to a typical TTL logic signal the RS-232 signal is much more immune to line noise and, therefore, is more likely to deliver a quality signal when transmitted long distances or in noisy environments.

Because of the noise immunity characteristic and the allowable voltage range (+3 to +15) and (−3 to −15), the RS-232 signal is capable of a much greater transmission distance than the TTL signal. RS-232 signals may be transmitted at least 50 feet and up to 100 feet depending upon the quality of the cable and characteristics of the electrical load at the receiving end of the cable.

Efforts to improve noise immunity and distance limitations of the RS-232 standard have lead to the development of other serial transmission standards. The RS-422 serial standard uses balanced differential amplifiers at both the transmitting and receiving ends of the cable. The use of differential amplifiers means that there are two wires per signal. This improved serial transmission standard allows cable lengths of up to 5,000 feet. Figure 3.13 provides a comparison between the primary communication interfaces that are used for the transmission of data (digital or analog) over distances beyond a few feet.

A reasonable generalization is that when digital signals need to be transmitted more than a few feet, the use of RS-232 or RS-422 serial transmission is a worthy consideration. RS-232 or RS-422 interface boards are available for PCs, minicomputers, and mainframe computers. To provide flexibility and to meet customers demands, producers of DA & C boards manufacture data acquisition and control boards ready to receive or transmit RS-232 signals, along with a variety of TTL I/O, or 4 to 20 mA inputs and outputs.

INPUT SIGNAL BUFFERING AND AMPLIFICATION

Some interfacing applications require impedance matching between the sensor and the input terminals on the DA & C board. The matching is needed so that output impedance of a sensor is matched with the input impedance of a DA & C board. Without close impedance matching a significant signal loss will occur.

Impedance matching can be accomplished by an operational amplifier connected as a unity gain amplifier. An impedance matching amplifier circuit is shown

Standard	Voltage	Transmission Range	Application
RS-232	+/-15	50 to 100 ft	Short Cables
RS-422	+5	5000 ft	Long Distance

Figure 3.13 Comparison of Transmission Standards

in Figure 3.14a. The gain from this form of circuit is one. A unity gain amplifier is not used because of its amplification qualities but rather for its high-input and low-output impedance characteristics. The high-input impedance of the operational amplifier minimizes the loading on the sensor and prevents signal degradation. The output impedance characteristic of the operational amplifier is ideal because it provides a low-output impedance for electrical loads such as meters, recorders, or connection to the DA & C board in a computer system.

Figure 3.14b shows an operational amplifier connected as an inverting amplifier. In this example, the circuit provides signal inversion, a gain of 10, and the high input, low output, impedance characteristic of operational amplifiers. Other amplifier gains can be obtained by changing the values of the two resistors shown in the schematic. This type of circuit is frequently used to amplify the sensor output before feeding it to the analog input of a data acquisition board.

CURRENT-TO-VOLTAGE CONVERSION

The conversion of current to voltage is frequently required in transducer circuits. The circuit shown in Figure 3.15 provides current to voltage (I to V) conversion. In this circuit, the current flowing through the light-dependent resistor is converted to voltage by the operational amplifier.

An added feature in this schematic is the inclusion of a provision to provide gain to the signal. The amplitude of the output voltage is controlled by the product of the current (I) and the value of the feedback resistor labeled (Rf).

ANALOG SCALING

Occasionally, analog signals produced by transducers are too large for direct connection to an analog-to-digital converter. The analog signal must be reduced (scaled) to a lower value prior to feeding the signal to the A/D circuitry on the interfacing board.

Amplitude reduction of the incoming analog signal can be performed by the classic voltage divider circuit shown in Figure 3.16. In this circuit the value of the voltage produced by the transducer can be *scaled* to any smaller value by selecting the proper ratio of resistors.

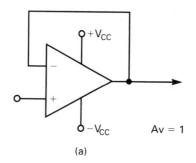

(a)

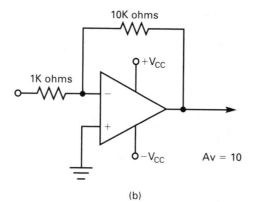

(b)

Figure 3.14 Signal Conditioning Using Operational Amplifiers

OFFSET VOLTAGE

The presence of an offset voltage is a very common problem associated with a variety of sensors and transducers. *Offset voltage* is an unwanted voltage that is produced by a sensor though the sensor theoretically should be producing zero volts.

An example of offset voltage can be found in the water tank problem described in Figure 3.8. That example presumes that when the water tank is empty the transducer produces 0 volts and when the water tank is filled the sensor would produce +5 volts. Unfortunately, calibration and manufacturing inaccuracies within the transducer do not permit this precise voltage output span from the transducer. In reality when the water tank is empty, the sensor is likely to produce approximately +1.0 volts. As the water level is increased, the voltage will rise proportionally until the maximum depth permitted by the range of the transducer is reached. When the water reaches the upper pressure limit of the sensor, the transducer will produce an output value of approximately +6 volts. In this example,

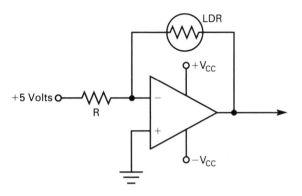

Figure 3.15 Current-to-Voltage Converter

the unwanted offset voltage is roughly $+1.0$ volts. Offset voltage is a voltage that is produced by the transducer even though the transducer is not measuring a meaningful physical variable such as pressure, temperature, and so on.

The Series 240PC family of pressure transducers manufactured by Micro Switch Corporation, 11 West Spring Street, Freeport, IL 61032 is typical of solid-state analog pressure transducers produced by a variety of manufacturers. The particular sensor, described in Figure 3.17 is the model 242PC05G which is rated at 0 to 5 pounds per square inch gauge (PSIG). In a water tank application, an upper limit rating of 5 PSIG equates to 138.4 inches of water.

The sensor is intended to provide an analog output voltage of 0 volts at atmospheric pressure and $+5$ volts at 5 PSI above atmospheric pressure. Unfortunately, this sensor produces an offset voltage of roughly $+1$ volt at atmospheric pressure. This means that at ambient atmospheric pressure the sensor produces an analog output of approximately $+1$ volt, when in theory the transducer should produce 0 volts. To insure accurate readings and a direct correlation between the value of the sensor's output and the applied pressure the $+1$ volt offset should be eliminated. To accomplish this necessary signal conditioning, the nulling circuit shown in Figure 3.18 may be used.

This circuit uses an operational amplifier connected in a summing application to eliminate the offset voltage. The summing circuit combines the $+1$ volt offset voltage with a -1 volt compensation voltage resulting in a "conditioned" output that can be fine tuned to achieve an output of 0 volts. Fine-tune nulling adjustment of the output voltage is accomplished by a *potentiometer*. After adjustment and calibration, the output signal level from the conditioning circuit will be 0 volts at atmospheric pressure. The linearity of the transducer and the span of the output voltage are not affected by the conditioning. Thus, at ambient pressure the analog output from the operational amplifier is 0 volts. Inclusion of the conditioning circuitry does not adversely effect the performance of the pressure sensor. As the

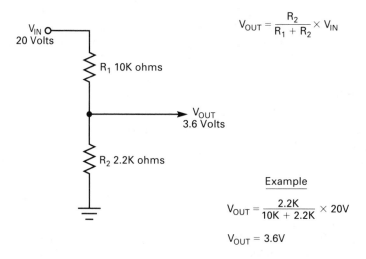

$$V_{OUT} = \frac{R_2}{R_1 + R_2} \times V_{IN}$$

V$_{IN}$ ○
20 Volts

R$_1$ 10K ohms

V$_{OUT}$
3.6 Volts

R$_2$ 2.2K ohms

Example

$$V_{OUT} = \frac{2.2K}{10K + 2.2K} \times 20V$$

$$V_{OUT} = 3.6V$$

Figure 3.16 Analog Scaling Circuit

pressure applied to the input of the pressure sensor increases toward the upper limit of 5 PSI, the analog output from the sensor increases toward +6 volts.

Other offset compensation circuits are available in addition to the schematic shown in Figure 3.18. All other forms of offset compensation circuits include a summing circuit and an operational amplifier. Another example of a compensation circuit is provided in Figure 3.19. This figure shows an inverting summing circuit. In this circuit, two input voltages are summed and inverted 180 degrees.

Figure 3.19 illustrates the resulting output voltage and polarity for three combinations of input voltage and different polarities. Notice that regardless of the amplitude or polarity of the incoming signals the conditioning circuit sums the two input voltages and produces and inverted resulting voltage value.

Operational amplifiers can also be used to provide differential amplification. Figure 3.20 shows a classic differential amplifier designed to receive inputs from two temperature transducers. The output signal from the differential circuit will be the difference between the two individual sensors. This type of circuit config-uration is ideal for monitoring the temperature defference between two locations.

The schematic shown in Figure 3.20 illustrates the use of two AD590 tem-perature transducers. These devices are produced by Analog Devices, One Tech-nology Way, Norwood, MA 02062. The AD590 provides a current in micro-amperes that is dependent upon the absolute temperature of the sensor. In this application, potentiometer R2 provides for elimination of any offset voltage. The circuit of Figure 3.20 illustrates a circuit that will amplify the difference between the tem-peratures at the location of the two sensors. This circuit employing differential amplification is useful when the difference between the two temperatures must be identified and amplified.

High level solid state pressure sensors   **240PC**

TYPICAL DIMENSIONS

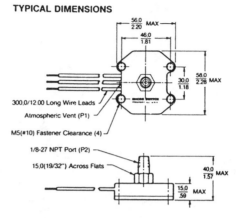

300,0/12.00 Long Wire Leads
Atmospheric Vent (P1)
M5(#10) Fastener Clearance (4)
1/8-27 NPT Port (P2)
15,0(19/32") Across Flats

FEATURES

- Compatible with wide range of non-caustic liquids
- Die-cast aluminum housing, epoxy painted
- Internal O-ring seals for contamination resistance
- Gage pressure measurement from ±2.5 psi to 0-250 psi
- Screw-in or flat-pack mounting
- 1/8-27 NPT port
- DC operation . . . linear, ripple-free output proportional to pressure
- Computer controlled laser trimming for close control of parameters
- Controlled null and full scale output, high performance temperature compensation
- Silicon sensor chip . . . integral diaphragm and ion implanted resistors
- High sensitivity, excellent stability
- 300 mm (12 in.) long, color coded #18 AWG leadwire termination

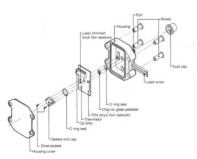

GENERAL SPECIFICATIONS*

Parameter	Min.	Typ.	Max.	Units
F.S.O. (Full Scale Output)**	4.85	5.00 ±2.5	5.15	V (1) (2)
Null Offset	0.95 3.45	1.00 3.50	1.05 3.55	V (1) (2)
Output at Full Pressure	5.80	6.00	6.20	V (3)
Response Time			1	msec
Excitation	7.00	8.00	16.00	VDC
Supply Current		8.00	20.00	mA

Output Current
  Source
  Sink
	10.0
	5.0
				mA

Ratiometricity 7 to 8 or 8 to 9V 8 to 12V		±0.50 ±2.00		%F.S.O.
Stability over 1 year		±1.00		%F.S.O.
Shock	MIL-STD-202, Method 213B (100G, half sine)			
Vibration	MIL-STD-202, Method 204C (10 to 2000 Hz at 10G)			

Temperature
  Compensated
  Operating
  Storage
	−18 to +63°C (0 to +145°F)
	−40 to +85°C (−40 to +185°F)
	−55 to +125°C (−67 to +257°F)			

Media Compatibility	Limited only to those non-caustic media which will not attack the diecast aluminum housing; silicon chip; glass; or one of the several O-ring materials (see Order Guide).			
Output Ripple	None, DC device			
Short Circuit Protection	Output may be shorted indefinitely to ground.			
Ground Reference	Supply and output are common.			

Figure 3.17 Micro Switch Series 240PC (242PC05G) Pressure Sensor
(Courtesy Micro Switch Corporation)

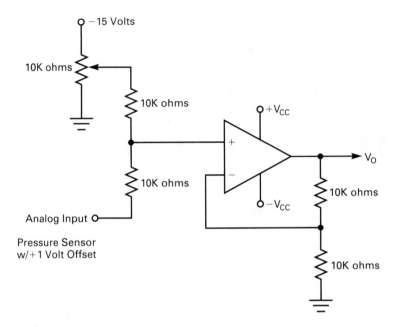

Figure 3.18 Offset Compensation Circuit

A variety of other temperature sensors are available and the reader is urged to consult data books produced by sensor manufacturers such as Analog Devices and National Semiconductor for more information. Another excellent source of information on a variety of sensors is the multi-volume set of catalogs produced by Omega Engineering, One Omega Drive, Stamford, CT 06907. These catalogs provide product descriptions, specifications, and detailed application information for a variety of sensors.

INTERFACING PROBLEMS

Problem #1: Wire the inverting operational amplifier shown in Figure 3.15. Apply a +1 volt DC voltage to the input pin of the amplifier and measure the amplifier's output voltage.

Apply a −1 volt DC voltage to the input pin of the amplifier and measure the amplifier's output voltage.

Problem #2: Construct the offset nulling circuit shown in Figure 3.18. Apply a +1.5 volt DC voltage to signal input. Adjust the compensation potentiometer until the "offset" voltage is eliminated.

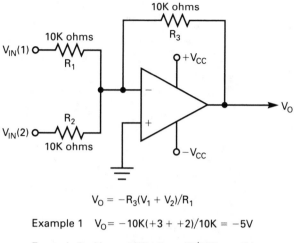

$$V_O = -R_3(V_1 + V_2)/R_1$$

Example 1 $V_O = -10K(+3 + +2)/10K = -5V$

Example 2 $V_O = -10K(+3 + -2)/10K = -1V$

Example 3 $V_O = -10K(-3 + +2)/10K = +1V$

Figure 3.19 Inverting Summing Circuit

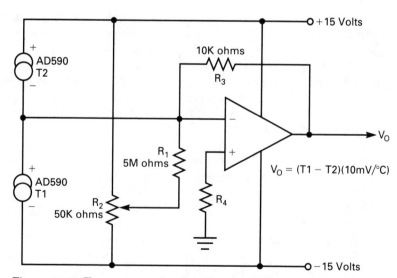

Figure 3.20 Temperature Application of a Differential Amplifier

CHAPTER 4

Time-of-Day Clock

TIME IN DATA ACQUISITION AND PROCESS CONTROL

Time is often an important variable in data acquisition and control applications. Time can be a factor in determining when data readings are taken, when machinery is turned on or off, and when processes begin and end. Several different techniques are used for measuring time. There are a variety of methods used to cause the computer to incorporate time in its decision-making process; all of which are available to technicians and engineers. When the computer uses time in the same context as humans use time (hours, minutes, and seconds), we will refer to this as *time-of-day*.

Everyday events occur with reference to time. We wake up for school or work at a particular time, perhaps 06:30, eat lunch at 12:00, and eat dinner at 18:00. These times refer to the 24-hour clock that is common to the military. Another aspect of time refers to the day of the week. It may be important for a computer which is controlling a system to distinguish between a work week (Monday through Friday) and weekends (Saturday and Sunday). With proper programming, the computer will "know" the time of day and also the day of the week.

An example of an application involving time-of-day together with a computer system can be illustrated by a security monitoring system that disables a store's burglar alarm during regular business hours, yet activates the alarm system during those hours that the store is closed. In this hypothetical time-of-day application, the computer needs to know the regular business hours of the store along with the actual time-of-day at any moment. If the current time is later than the time the business was scheduled to close, the computer arms the security system, activating motion sensors in the store. If one of the sensors were to detect a burglary in progress, and the time-of-day is later than closing time, the burglar alarm bell would be turned on. Conversely, if the time-of-day were within normal business hours, the alarm system would be disabled. The decision to arm the burglar alarm system and the resulting action if a burglary is detected is tied to *absolute time-of-day*.

GETTING THE TIME-OF-DAY

Every personal computer has a built-in clock that is capable of monitoring the time-of-day. The time-of-day clock is evident to computer users in the form of *date* and *time stamping* of files which are then saved to a disk. Every time a file is saved to a disk, the time and date are stored as part of the directory entry.

The format and character of a computer time-of-day clock may take different forms. Some PCs have a hardware clock that is battery powered and keeps track of the time and date even when the computer is off. Other computers have clocks that need to be told the time and date each time the computer is booted. In these computers, the user is prompted to enter the time and date during the power-up sequence. Whatever the method of setting the time and date, every PC has an internal clock that keeps the time-of-day. Using software together with the internal time-of-day, the time and date can be accessed and used to control the acquisition of data or control of external devices.

ACCESSING TIME-OF-DAY WITH THE C LANGUAGE

The ANSI C language standard includes several functions for determining time of day. In the example shown in Figure 4.1, the date, day of week, and time are printed on the CRT screen.

The function, **time()**, returns in **t** the number of seconds that have elapsed since 00:00:00 Coordinated Universal Time January 1, 1970. The function, **localtime()**, converts **t** to a structure of date, hours, minutes, and seconds. This structure assumes the following form:

```
int tm_sec;      seconds after the minute (0..59)
int tm_min;      minutes after the hour (0..59)
int tm_hour;     hours since midnight (0..23)
int tm_mday;     day of the month (1..31)
int tm_mon;      month since January (0..11)
int tm_year;     year since 1900
int tm_wday      day of week since Sunday (0..6)
int tm_yday;     day of year since January 1 (0..365)
int tm_isdst;    Daylight Savings Time flag
```

```
/*                                                                    */

#include <stdio.h>
#include <time.h>

main()
  {
    struct tm *local;              /* C time/date structure    */
    time_t t;                      /* equals the number of
                                      seconds since 1/1/70      */

    t=time(NULL);                  /* get the seconds           */
    local=localtime(&t);           /* convert to local          */
    printf("%s",asctime(local));   /* print as time string      */
    return 0;                      /* always a good idea to
                                      return a success code     */

  }
```

Figure 4.1 Reading Time-of-Day Clock using C

Figure 4.1 uses **asctime()** to convert this information to a string. If a complete string is not needed, specific data such as day of the year can be isolated from the structure.

Figure 4.2 illustrates the use of part of the time structure. The C program shown in Figure 4.2 determines the number of days remaining until Christmas. This program uses the **int tm_yday** portion of the structure. The day of the year is subtracted from 358 yielding the number of days remaining until Christmas.

In the example shown in Figure 4.2, **local** points to a structure that has **tm_yday** as one of its fields as prescribed by the ANSI standard. The **printf** statement prints 358 minus the number of days since January first. This technique is typical of the procedure used to access individual elements of the standard.

ACCESSING SYSTEM TIME WITH BASIC

The BASIC language can also be used to access the system time and date. Although BASIC does not provide the day of the week and day of the year information, it does return time information in string format. Figure 4.3 shows a program segment that assigns the current time to the string variable **A$** and the date to **B$**.

Figure 4.3 shows the simplicity of accessing the system time and date using the BASIC language. The built-in functions **TIME$** and **DATE$** provide the numeric-to-ASCII conversion. Though the time information is in the form of a string, specific parts of the string can still be accessed. For example, if only seconds were to be accessed, the code segment shown in Figure 4.4 could be used.

The variable **SEC** contains the decimal value of the seconds rather than the ASCII string. Providing seconds in a decimal format means that mathematics functions can now be performed on the portion of the string that has been isolated.

```
/*                                                                      */
#include <stdio.h>
#include <time.h>
#include <conio.h>

main()
  {
     struct tm *local;              /* C time/date structure    */
     time_t t;                      /* equals the number of
                                       seconds since 1/1/70      */

     t=time(NULL);                  /* get the seconds           */
     local=localtime(&t);           /* convert to local          */
     gotoxy(10,10);                 /* set screen position       */
     printf("There are %d shopping days left till Christmas",
                 (358-local->tm_yday));  /* print as days till... string */
     return 0;                      /* always a good idea to
                                       return a success code     */
  }
```

Figure 4.2 Accessing Individual Time-of-Day Elements using C

ACCESSING SYSTEM TIME WITH OPERATING SYSTEM CALLS

Both languages, C and BASIC, offer programmers another method of getting the time-of-day. As indicated in Chapter 2, all system resources are available through interrupts. The programmer can invoke a software interrupt to get the time of day. Refer to your C manual for information on the **INTDOS** function and BASIC's counterpart, the **CALL INT86OLD** command. Both commands require, as parameters, an input register array and a return register array. Additionally, BASIC's CALL INT86OLD needs the interrupt number (21 hex for all DOS functions). This parameter is implied in the C function.

When one of these functions is used, the return register array contains time or date values depending on which function is called. The following list contains the registers and their values upon returning from these function calls.

Time	CH = Hours (0–23)
	CL = Minutes (0–59)
	DH = Seconds (0–59)
	DL = hundredths of a second (0–99)
Date	CX = Year
	DH = Month (1 = January, 2 = February, etc.)
	DL = Day of Month (1–31)
	AL = Day of Week (0 = Sunday, 1 = Monday, etc.)

Figure 4.5 provides an example of this call using the C language. The same program written in BASIC is shown in Figure 4.6. The DOS function number for getting the time of day is 2C00 hex and the date is 2A00 hex.

```
,
,
,
,
,
A$ = TIME$                 'Get the time
B$ = DATE$                 'Get the date
LOCATE 10, 10              'Set the screen position
PRINT B$                   'Print the date
LOCATE 11, 10              'Set the new screen position
PRINT A$                   'Print the time
```

Figure 4.3 Time-of-Day using BASIC

SETTING THE SYSTEM TIME AND DATE

Occasions may arise when the programmer wants to change the current value for the system time-of-day. To set the time-of-day or the date from BASIC, a simple assignment statement such as **TIME$** = **"15:33:45"** may be used. The equivalent assignment statement for the date is **DATE$** = **"12/25/1989"**. The only restriction is that the programmer will not be allowed to set illegal values, such as an hour of 25 o'clock or a month identified by the number 13.

Setting the time in C language requires the use of DOS function calls. While the DOS function **0x2C** returns the system time in the return register array, function **0x2D** expects the new time values in the entry register array. Similarly, changing the system date can be accomplished by loading the entry array with the new date values and calling DOS function **0x2B**. In either case, the entry array has the same form as the return array. That is, the year is in CX, the month in DH, and so on. Figure 4.7 provides a listing of the real-time entry array.

USING THE TIME-OF-DAY CLOCK

When a program is written that asks the system for the time-of-day, the system responds with an absolute measure of time. This absolute value is important in the decision-making process that is established by the software being executed. The burglar alarm system described in the beginning of this chapter is totally dependent upon time-of-day to figure out if the alarm should be activated or deactivated. Other time-oriented applications do not require an absolute value but rather a relative measure of elapsed time.

An example of relative time can be provided by the process of preparing a hard-boiled egg. For example, cooking a three-minute egg is not influenced by the actual time-of-day; but it is influenced by the amount of time that the egg has been in the boiling water. In this hard-boiled egg example, a measure of relative time can be found by recording the START time and adding the LENGTH of the process (three minutes) to figure out the STOP time. This process uses a software program

```
'
'
'
'
'
A$ = TIME$                  'Get the time
LOCATE 10, 10               'Set the screen position
SECOND$ = RIGHT$(A$, 2)     'Isolate the secnds
SEC = VAL(SECOND$)          'Convert string to Decimal
PRINT SECOND$               'Print the seconds as a string
LOCATE 11, 10               'Set the new screen position
PRINT SEC                   'Print it again as a number
```

Figure 4.4 Accessing Seconds using BASIC

that is continually polling the current time and comparing it to the STOP time. When the current time and the STOP time are equal, the hard-boiled egg is presumed to be cooked.

This technique is an effective method of determining relative time; however, the disadvantage of this method is that the computer is committed to reading the current time continually and comparing that current time with the calculated STOP time. This process is accomplished in a *software loop*. If the computer is "tied up" in the loop, it cannot do other tasks or control other processes that are outside of the loop. If, in response to this dilemma, the programmer includes a large number of tasks in the loop, the extended time it takes to go through the loop might cause the proper STOP time to be missed and the eggs would be over cooked. This dilemma is a real consideration when developing process control programs.

We will present another way to find relative time in Chapter 11. That method of monitoring relative time uses another DA & C board so that the computer can do other tasks but allows the process underway to be stopped when the proper time has arrived.

INTERFACING PROBLEMS

Problem #1: Write a program in C or BASIC that will read the system clock and continuously print the time at the same place on the screen. The format for the time string should be HH:MM:SS where HH is the hours, MM is the minutes, and SS is the seconds.

Problem #2: Write a program so that the computer will print "Hello World!" on the CRT screen at a location of your choice other than (0,0). After the message has been displayed for five seconds, the message should disappear and the screen should be blank. After another five seconds has elapsed, the computer should again print the message "Hello World" on the screen. The cycle composed of the five second intervals should continue.

```
/*                                                              */


#include <stdio.h>
#include <dos.h>
#include <conio.h>

#define TIME   0x2C00            /* DOS Time Function #      */
#define DATE   0x2A00            /* DOS Date Function #      */
main()

   {
     union REGS regs;           /* Set up space for calls   */

     regs.x.ax = TIME;          /* Get the time             */
     intdos(&regs, &regs);                  /* Make the call using the same
                                register buffers for the
                                call and return values    */
     clrscr();                  /* Clear the screen         */
     gotoxy(35,10);             /* Set the screen location  */
     printf("The seconds are %d",regs.h.dh);/*Print the seconds         */

     regs.x.ax = DATE;          /* Get the Date             */
     intdos(&regs, &regs);      /* Make the call using the same
                                register buffers for the
                                call and return values    */
     gotoxy(35,11);             /* Set the screen location  */
     printf("The day is ");
     switch (regs.h.al)
        {
          case (0): `
            printf("Sunday");
            break;
        }
   }
```

Figure 4.5 DOS Call using C

Problem #3: Write a program which provides the computer user the current system time and date. Prompt the computer user if he wants to change either the time or the date. If the user does wish to change either variable, allow for the input of the new values and then check them for proper limits. For example:

Month	1–12
Day	1–31
Year	1980–2099
Hour	0–23
Minute	0–59
Second	0–59

If the new values are within limits, allow the software to set the system time and/or date and *report the new value to the user.*

```
'
' It is importyant to remember to start Quick Basic with the line:
'              qb /lqb.qlb
' This loads the Quick library QB.QLB necessary for the IN86OLD function
'
'
'$INCLUDE: 'QB.BI'                        'Include file
DIM inary%(7), outary%(7)                 'Define arrays for DOS call

CONST ax = 0, bx = 1, cx = 2, dx = 3, bp = 4, si = 5, di = 6, FL = 7
                'These are used to point to the approprite members

inary%(ax) = &H2C00                       'DOS time Function #
CALL INT86OLD(&H21, inary%(), outary%())'Call DOS Function
CLS
LOCATE 10, 35
PRINT "The seconds are "; INT(outary%(dx) / 256)

inary%(ax) = &H2A00                       'DOS date Function #
CALL INT86OLD(&H21, inary%(), outary%())'Call Dos Function
LOCATE 11, 35
PRINT "The day of the week is ";
SELECT CASE (outary%(ax) AND 255)
        CASE 0
            PRINT "Sunday"
        CASE 1
            PRINT "Monday"
        CASE 2
            PRINT "Tuesday"
        CASE 3
            PRINT "Wednesday"
        CASE 4
            PRINT "Thursday"
        CASE 5
            PRINT "Friday"
        CASE 6
            PRINT "Saturday"
END SELECT
```
Figure 4.6 DOS Call using BASIC

```
Time        CH - Hours (0-23)
            CL - Minutes (0-59)
            DH - Seconds (0-59)
            DL - Hundredths of a Second (0-99)

Date        CX - Year
            DH - Month (1=January, 2=February, etc.)
            DL - Day of Month (1-31)
            AL - Day of Week (0=Sunday, 1=Monday, etc.)
```
Figure 4.7 Time-of-Day Array

CHAPTER **5**

Interfacing to the PC Bus: Digital Input

Digital input is the simplest form of inputting signals created by external devices to a computer. *Digital output* is the simplest form of outputting from a computer to external control devices. *Digital input/output* (I/O) provides for the inputting of digital logic signals to the computer from the "real world" and the outputting of compatible logic signals to the "real word." This form of interfacing uses a digital I/O data acquisition and control board to connect the PC bus with input devices such as switches or other sensors. The DA & C I/O board also connects the computer's bus to electrical loads such as relays, indicator lights, or other "real-world" electrical devices.

Digital interface boards are manufactured by many vendors. Regardless of the manufacturer, these boards plug into the computer's bus and have similar electrical characteristics and specifications. Since this text is oriented toward the IBM-PC, the interface boards described will be IBM-PC compatible. Other computer busses have I/O boards that are designed to be compatible with those busses.

The PC compatible I/O board provides the necessary interface between the computer bus and the wires going to and coming from the external real-world devices. Typical of many digital I/O boards available on the market today is the Model PIO-12, which is manufactured by Keithley MetraByte Corporation, 440 Myles Standish Boulevard, Taunton, MA 02780. This chapter will focus on this I/O board because it is typical of several similar digital I/O boards. The Model PIO-12 was selected because of its low cost, availability, and popularity. The PIO-

12, pictured in Figure 5.1, plugs into the 62 pin IBM-PC, XT, or AT bus and provides three 8-bit bidirectional I/O ports. The designation of these ports as either input or output ports is easily accomplished through software commands.

Other manufacturers producing I/O boards compatible with the PIO-12 include:

- Computer Boards, Inc., 44 Wood Avenue, Mansfield, MA 02048;

- Industrial Computer Source, 10180 Scripps Ranch Boulevard., San Diego, CA 92131; and

- CyberResearch, Inc., 25 Business Park Drive, Branford, CT 06405.

These compatible boards are both software and hardware interchangeable with the PIO-12.

The three I/O ports (designated by the alphabetic letters A, B, and C) may be configured in a variety of input and output combinations. For example, in the simplest configuration, all three ports are configured as *outputs* from the computer or all three ports may be configured as *inputs* to the computer.

The PIO-12 I/O board may also be configured so that one or two of the ports serve as either input or output ports, while the third port is used in a *handshaking* role. Handshaking provides digital signaling communication between the computer and external sensors or between load devices and the computer. Handshaking provides a method through which the computer is aware of the status of connected external devices. For example, through handshaking the computer can be informed that a connected printer is not turned on or perhaps that the printer has run out of paper. Handshaking can also inform the computer that a sensing device is not functioning properly or that an object on a conveyor belt is improperly positioned and that it is likely to cause damage to the material processing system. With proper design and forethought, handshaking can play an important role in data acquisition and control systems.

Computers that have the IBM MicroChannel bus (such as the IBM PS-2 Models 50 through 80) require I/O boards that are different from those used in the traditional IBM-PC, XT, or AT bus. The MetraByte PIO-12 will not operate in the MicroChannel bus. If a MicroChannel computer is used in an application requiring digital I/O, the MetraByte uCPIO-12 I/O board should be substituted for the PIO-12. This board is compatible with the MicroChannel architecture and must be used in computers using this alternate bus structure. Figure 5.2 provides additional information about the uCPIO-12 board.

BASE ADDRESS SELECTION

Before a digital I/O board like the PIO-12 can be installed into a personal computer, the port address of the I/O board must be selected. The selection of this base address is essential because each I/O device connected to the computer must have

FEATURES

- 24 TTL/DTL digital I/O lines
- ±12 V, ±5 V power from IBM PC/XT
- Unidirectional, bidirectional and strobed I/O
- Interrupt handling
- Direct interface to wide range of accessory products
- Plugs into IBM PC/XT/AT bus
- Handshaking
- Compatible with ERB-24, SSIO-24, SRA-01 and ERA-01

APPLICATIONS

- Contact closure monitoring
- Digital I/O control
- Useful with A/Ds and D/As
- Communication with other computers
- Operate relays (PIO-24 recommended)
- Alarm monitoring

FUNCTIONAL DESCRIPTION

Keithley MetraByte's PIO-12 board provides 24 TTL/DTL compatible digital I/O lines. It is a flexible interface for parallel input/output devices such as instruments and displays and user constructed systems and equipment.

Twenty-four digital I/O lines are provided through an 8255-5 programmable peripheral interface (PPI) IC and consist of three ports: an 8-bit PA port, an 8-bit PB port, and an 8-bit PC port. The PC port may also be used as two half ports of 4 bits, PC upper (PC 4 – 7) and PC lower (PC 0 – 3). Each of the ports and half ports may be configured as an input or an output by software control according to the contents of a write only control register in the PPI. The ports may be both read and written. In addition, other configurations are possible for unidirectional and bidirectional strobed I/O where the PC ports are used for control of data transfer

and interrupt generation, etc. Users are referred to the Intel 8255-5 data sheet for a complete technical description and summary of the various operating modes of the PPI.

Interrupt handling is via a tristate driver with separate enable (interrupt enable — active low). This may be connected to interrupt levels 2 – 7 on the IBM PC bus by a plug-type jumper on the board. Handling of an interrupt is controlled by the 8259 interrupt controller in the IBM PC and this is set by BIOS on system initialization to respond to positive (low – high) edge triggered inputs. Users must program the 8259 to respond to their requirements and set up corresponding interrupt handlers. Interrupt input and enable lines and external connections to the PC's bus power supplies (+5V, +12 V, -12 V and -5 V) are available at the connector.

SPECIFICATIONS

Logic Inputs and Outputs	Min	Max
Input logic low voltage	–0.5 V	0.8 V
Input logic high voltage	2.0 V	5.0 V
Input load current PA, PB, PC port		
(0 < V in < 5 V)	–10 µA	+10 µA
Input low current, interrupt inputs	—	–0.4 mA
Input high current, interrupt inputs	—	20 µA
Output low voltage PA, PB, PC ports		
(Isink = 1.7 mA)	—	0.45 V
Output high voltage PA, PB, PC ports		
(Isource = 200 µA)	2.4 V	

All outputs and inputs are TTL/DTL-compatible and outputs will drive one standard TTL load (74 series) or 4 LSTTL (74LS) loads. CMOS compatibility can be obtained by connecting a 10 k ohm pull-up resistor from the input or output to +5 V.

Outputs do not have the drive current to control most solid state relays directly (e.g., OACs, ODCs). Interface boards such as the SRA-01 and SSIO-24 provide buffers which allow the PIO-12 to control these relays.

Power requirements

	+5 V	170 mA typ
	+12 V	not used
	–12 V	not used

Environmental

Operating temperature range	0 to 50 °C
Storage temperature range	-40 to + 100 °C
Humidity	0 to 90% non-condensing
Dimension	5 in x 4.5 in x .75 in (half slot) (12.7 cm x 10.8 cm)
Weight	4 oz (113 g)

BLOCK DIAGRAM

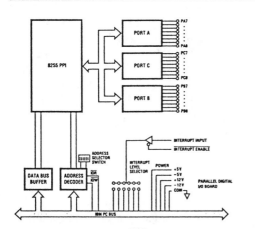

Figure 5.1 MetraByte PIO-12 Digital I/O Board *(Courtesy Keithley MetraByte Corporation)*

FEATURES

- IBM PS/2 Models 50 – 80 compatible
- 24 TTL/CMOS compatible I/O lines
- Unidirectional, bidirectional or strobed I/O
- Complete interrupt interface circuitry
- Compatible with ERB-24, ERA-01, SRA-01, SSIO-24

- Brings out ±12 V, +5 V from PS/2
- Includes complete user's guide and installation instructions

APPLICATIONS

- Contact closure monitors
- BCD interfaces
- Digital I/O control
- External circuitry interface
- Printer interfaces

FUNCTIONAL DESCRIPTION

Keithley MetraByte's μCPIO-12 24-bit parallel digital I/O board provides a simple and inexpensive means of interfacing an IBM PS/2 Model 50, 60, 80 or other Micro Channel compatible computers to a wide variety of digital applications. In addition to the 24 TTL/NMOS/CMOS-compatible data lines, the μCPIO-12 offers full access to the PS/2's interrupt lines and allows external connection to the computer's power supplies. For applications with non logic-level signals, Keithley MetraByte offers the following signal conditioning boards: SRA-01 8-channel Solid State I/O module board, ERA-01 8-channel electromechanical relay board, ERB-24 24-channel electromechanical relay board and SSIO-24 24-channel solid state I/O module rack.

The 24 digital I/O lines are provided by an Intel 8255 programmable peripheral interface. The 8255 divides the 24 bits into three separate 8-bit ports (PA, PB and PC) that can be set independently as inputs or outputs. In addition to operating as a standard 8-bit data port, the third port (PC) is subdivided into two 4-bit ports or can provide handshaking signals for the other two 8-bit ports.

In keeping with the design specifications for the PS/2 adapter cards, there are no dip switches or user adjustments on the μCPIO-12. Base address and interrupt level selection are performed by the PS/2 set-up program. IBM has allocated board identification number 6028 (hex) to the μCPIO-12. Along with the μCPIO-12, Keithley MetraByte supplies a complete user's manual and a 3½-inch disk with all required installation and operation software. The board has been designed to plug into any available Micro Channel expansion slot within the computer.

The μCPIO-12 is easily configured to interface to BCD equipment, monitor switches and contact closures, control relays, etc. Applications for the board include equipment control, keypad scanning, printer interface, motor control, intruder alarms, energy management and interface to custom external circuitry.

BLOCK DIAGRAM

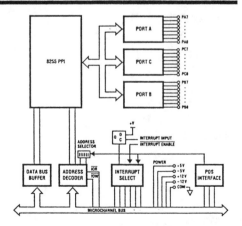

Figure 5.2 IBM MicroChannel Digital I/O Board *(Courtesy Keithley MetraByte Corporation)*

a unique I/O address designation. The typical IBM-PC bus provides for 1024 I/O port address locations. The range of addresses is encoded by 10 address bits that are used to identify which of the I/O ports is active.

Many of the 1024 addresses are permanently assigned to specific functions in the computer system. Figure 5.3 illustrates an I/O port map for the IBM-PC. This figure shows that addresses 000 Hex through 1FF Hex are assigned to the PC's systems board. These 512 (000-1FF) addresses provide for functions such as interrupts, DMA operations, and other peripheral interfaces to the computer.

Hex range	Usage	
000-00F	DMA chip 8237A-5	
020-021	Interrupt 8259A	
040-043	Timer 8253-5	
060-063	PPI 8255A-5	Assigned to
080-083	DMA page registers	system board
0Ax	NMI mask register	components
0Cx	Reserved	
0Ex	Reserved	
100-1FF	Not usable	
200-20F	Game control	
210-217	Expansion unit	
220-24F	Reserved	
278-27F	Reserved	
2F0-2F7	Reserved	
2F8-2FF	Asynchronous communications (2)	
300-31F	Prototype card	
320-32F	Fixed disk	
378-37F	Printer	Assigned to
380-38C	SDLC communications	feature card
380-389	Binary synchronous communications (2)	ports
3A0-3A9	Binary synchronous communications (1)	
3B0-3BF	IBM monochrome display/printer	
3C0-3CF	Reserved	
3D0-3DF	Color/graphics	
3E0-3F7	Reserved	
3F0-3F7	Diskette	
3F8-3FF	Asynchronous communications (1)	

Figure 5.3 I/O Port Map *(From Tompkins & Webster;* Interfacing Sensors to the IBM PC, *Prentice-Hall, 1988)*

The 512 address locations that range from 200 Hex through 3FF Hex are assigned to a wide range of accessory boards that typically are plugged into the computer bus connector. Boards such as a floppy-disk controller, hard-disk controller, printer board, or color/graphics boards often use these addresses. Designers of the PC provided a minimum of 32 port addresses that are not designated for other applications and, therefore, can be used with *user-added boards*. These 32 available ports have Hex addresses that range from 300 Hex through 31F Hex and ideally are suited for use by DA & C I/O boards.

I/O boards such as the MetraByte PIO-12 typically are addressed within this range of port addresses. Addresses outside the 300-31F range can be used for I/O boards, however, addressing conflicts and operational difficulties may arise if the addresses are not cho...n with the greatest of care and forethought! If addresses for I/O ports are selected that fall out of the 300 Hex-31F Hex range, address conflicts must be avoided.

Setting of the base address for each I/O board installed in the computer is accomplished by a DIP switch on the board. The 10-bit base address is encoded by an eight position DIP switch. It may seem contradictory that an 8-bit DIP switch is used to encode a 10-bit address. However, this is not as confusing as it appears since only the upper eight bits of the address needs to be set with the DIP switch.

The reason that only eight bits of the address need to be decoded is based upon the concept that each PIO-12 board has three I/O ports and one control register. Each of the ports and the control register occupy a separate port address, thus four sequential port addresses are used or "occupied" by each PIO-12 I/O board. Realizing that four addresses can be encoded by two bits (00 Hex-11 Hex) helps explain why only eight of the 10 address bits needs to be decoded to establish the *base address* of the I/O board. Figure 5.4 shows the eight position DIP switch as found on the PIO-12 board.

Notice that the upper eight bits of the addresses (A9-A2) are set by the DIP switch. Thus, the I/O board can be set to a base address in increments of every four address locations. Typical base addresses for the PIO-12 board installed in a PC, XT, or AT are 300 Hex, 304 Hex, 308 Hex, 30C Hex, 310 Hex, 314 Hex, and so on.

INPUT OR OUTPUT SELECTION

Designation of whether an I/O port is configured for input or output is determined by the software. The PIO-12 board assigns the three I/O ports to the first three addresses beginning with the base address. The fourth address is assigned to the control register on the board. The control register retains the binary code that determines the input or output configuration of the three I/O ports. This arrangement, one control register and three I/O ports, accounts for the requirement that four port addresses be assigned to each PIO-12 board. Figure 5.4 illustrates the port designation and control register for a typical PIO-12 board having a base

Specifying the Base Address

The Base Address switch is preset at the factory for 300-Hex.
If this address is already assigned to some other device in
your computer, you must set the Base Address switch to specify
a different address.

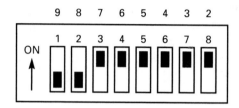

BASE ADDRESS

Figure 5.4 Address Selection DIP Switch

		Value when Switch is OFF	
Switch	Address line	Decimal	Hex
1	A9	512	200
2	A8	256	100
3	A7	128	80
4	A6	64	40
5	A5	32	20
6	A4	16	10
7	A3	8	8
8	A2	4	4

address of 300 Hex. Observe that Port A is represented by the address 300 Hex,
that Port B is identified by address 301 Hex, Port C is identified by address 302
Hex, and the control register for this I/O board is addressed as 303 Hex. This
arrangement is typical for the PIO-12 and similar DA & C boards that use the
popular 8255 interfacing integrated circuit chip.

Programming of the control register on the I/O board can be accomplished by using any computer programming language including C, BASIC,
or Assembly Language programming. As mentioned earlier, this text will be

providing programming examples in C and BASIC. Sample programming statements will be provided to illustrate how the three ports can be configured on the PIO-12 board.

The primary integrated circuit on the PIO-12 is a chip produced by the INTEL Corporation, identified as the 8255 *programmable peripheral interface* (PPI) *chip*. This device is very popular as a multi-port interface chip and is widely used in a variety of computer applications in products other than the PIO-12 board.

To understand I/O capabilities and handshaking options available to the user of the PIO-12 board, it is necessary to understand the Intel 8255 chip. A simplified block diagram of the 8255 PPI chip is shown in Figure 5.5.

The drawing shows the three I/O ports (A, B, and C) along with the control register. The 8255 chip has three operational modes identified as *mode 0*, *mode 1*, and *mode 2*. Each mode defines a particular configuration of the ports such as the input, output, and handshaking role that the three I/O ports can play. Mode 0 is the basic I/O configuration for the 8255 and PIO-12 board. In this mode, the 8255 chip provides two 8-bit ports (A and B) and two 4-bit ports (identified as upper portion of port C and lower portion of port C). The two parts of port C can be thought of as a complete 8-bit port that is available in two halves.

In mode 0 each of the three ports can be configured into any combination of input and output ports that the programmer prefers. In mode 0, the basic I/O mode, the output bits on any port are latched while the input bits are tri-stated. The following three control word examples provide illustrations of a few of the I/O combinations available when using the 8255.

Control byte: 10000000 = 80 Hex = A, B, and C output ports

Control byte: 10011011 = 9B Hex = A, B, and C input ports

Control byte: 10010010 = 92 Hex = A and B input, C output

The first control byte uses mode 0 and establishes all three ports as output ports. The second example establishes ports A, B, and C as input ports. The third example illustrates the control word necessary to configure A and B as input ports and port C as an output port. Mode 0 is a very straight forward, easy-to-use mode that can provide for a total of 16 different I/O configurations.

Figure 5.6 illustrates the choice of mode definition formats available for the 8255 chip. Mode 1, the strobed I/O mode, provides two 8-bit ports that may be configured as either input or output. Two 4-bit portions of port C become handshaking ports that work with the A and B ports. Thus, port C gives up its role as an I/O port and supports the other two existing ports in a handshaking capacity. The upper portion of port C (bits D3-D7) serve handshaking functions associated with port A. The lower portion of port C (bits D0-D2) serves a handshaking role for port B.

Each handshaking bit assumes a different function depending upon whether that bit is serving as support to an output port or an input port. Figure 5.7 illustrates

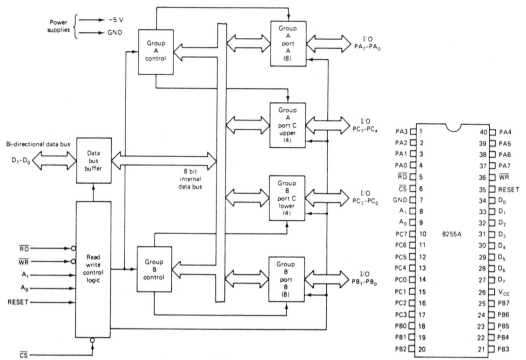

Figure 5.5 8255 P.P.I. Chip Architecture *(From Singh & Triebel, The 8086 and 80286 Microprocessors, Prentice-Hall, 1990 Courtesy Intel Corporation)*

two different I/O configurations with the appropriate handshaking role performed by port C.

The first example in Figure 5.7(a) shows both ports, A and B, as input ports. Notice that bit 4 and bit 2 on port C serve as strobe inputs for their respective 8-bit ports. When the appropriate strobe pin is brought low (0 volts), data will be loaded into the 8-bit port. "Strobing" bit PC4 or PC2 will latch the data into the corresponding input port. In response to the latching of the data, the *input buffer full* (IBF) *bit*, PC5 or PC1, will go high (logic 1). This bit serves as an acknowledgment that the data was latched and is available at the appropriate input port. Each half of port C also provides an *interrupt request* (INTR) *bit*, PC3 and PC0. These bits will go high (logic 1) when an external device is requesting service. These bits may be used to interrupt the CPU and can be a convenient method of allowing an input device to request service from the computer by simply strobing the input port.

Figure 5.7(b) illustrates how ports A and B may be configured as output ports along with the appropriate handshaking bits from port C. In this configuration, bits 3, 6, and 7 from port C serve new roles. Bit PC7 serves the role of an *output buffer full* (OBF) *indicator*. In this role, the OBF bit goes low (logic 0) to show

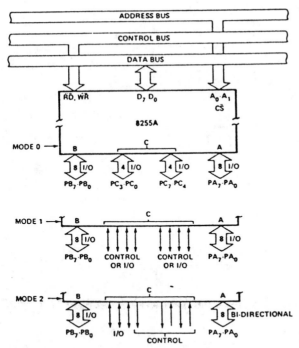

Figure 5.6 Mode Definition Examples *(Courtesy Intel Corporation)*

that the output port has had 8-bits of data written to it and it is holding the data in an output latch. Bit 6 (PC6) serves as an *acknowledge input* (ACK) from the output device. The output device acknowledges receipt of the data by applying a logic 0 to the ACK bit. The acknowledgment pulse clears the OBF bit and allows the data outputting process to the buffer to be repeated.

For simplicity, the examples provided in Figures 5.7(a) and 5.7(b) show both ports as either input or output ports. This need not be the case. Both ports do not necessarily need to be configured the same. In mode 1, port A and port B can be configured for different data directions. Thus port A could be programmed for an input port with appropriate handshaking and port B could be programmed for output with the appropriate handshaking. Mode 1 also supports the reversal of the I/O roles of port A and port B.

Mode 2 is the least likely used mode. In this mode, the chip establishes the eight bits of port A as a bidirectional port, allowing data to be passed in both directions on the same wires. Port B is not used in this mode. However, five bits in port C serve as bidirectional handshaking bits.

No matter which mode is used and regardless of which combination of input or output ports is wanted, the control byte (eight bits) is always written to the control register in the 8255 chip. The control register is addressed as the base

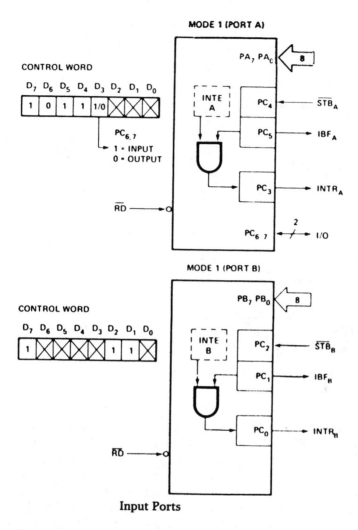

Figure 5.7(a) Mode 1 Examples *(Courtesy Intel Corporation)*

address of the board plus 3. For a MetraByte PIO-12 or equivalent I/O board with a base address of 300 Hex the address of the control register is 303 Hex. Regardless of the programming language used, writing a control byte to the 303 Hex address will configure the I/O board.

Figure 5.8 provides a *mode definition summary* for the 8255 PPI chip. This summary illustrates the capabilities of each I/O port and identifies the handshaking role of each bit in port C.

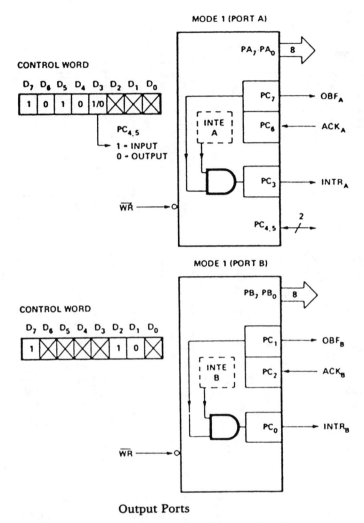

Figure 5.7(b) Mode 1 Examples *(Courtesy Intel Corporation)*

It is important to realize that while the base address of the PIO-12 I/O board is permanently set via the DIP switch, the mode and data direction selection of the I/O ports is not permanent and may be changed by software at any time.

Each time the computer is turned ON or the computer is **Reset** the configuration code must be written to the PPI control register. Because of this characteristic, and for software documentation reasons, the control byte should be placed

Mode Definition Summary

	MODE 0 IN	MODE 0 OUT	MODE 1 IN	MODE 1 OUT	MODE 2 GROUP A ONLY
PA_0	IN	OUT	IN	OUT	← →
PA_1	IN	OUT	IN	OUT	← →
PA_2	IN	OUT	IN	OUT	← →
PA_3	IN	OUT	IN	OUT	← →
PA_4	IN	OUT	IN	OUT	← →
PA_5	IN	OUT	IN	OUT	← →
PA_6	IN	OUT	IN	OUT	← →
PA_7	IN	OUT	IN	OUT	← →
PB_0	IN	OUT	IN	OUT	——
PB_1	IN	OUT	IN	OUT	——
PB_2	IN	OUT	IN	OUT	——
PB_3	IN	OUT	IN	OUT	——
PB_4	IN	OUT	IN	OUT	——
PB_5	IN	OUT	IN	OUT	——
PB_6	IN	OUT	IN	OUT	——
PB_7	IN	OUT	IN	OUT	——
PC_0	IN	OUT	$INTR_B$	$INTR_B$	I/O
PC_1	IN	OUT	IBF_B	$\overline{OBF}_B$	I/O
PC_2	IN	OUT	$\overline{STB}_B$	$\overline{ACK}_B$	I/O
PC_3	IN	OUT	$INTR_A$	$INTR_A$	$INTR_A$
PC_4	IN	OUT	$\overline{STB}_A$	I/O	$\overline{STB}_A$
PC_5	IN	OUT	IBF_A	I/O	IBF_A
PC_6	IN	OUT	I/O	$\overline{ACK}_A$	$\overline{ACK}_A$
PC_7	IN	OUT	I/O	$\overline{OBF}_A$	$\overline{OBF}_A$

MODE 0 OR MODE 1 ONLY

Figure 5.8 Intel 8255 Mode Definition Summary *(Courtesy Intel Corporation)*

in the beginning of an I/O program. The writing of a data byte to the control register should be identified as a *port initialization instruction* in the comment section of the program.

On power up, or after a *reset* of the microprocessor, the I/O ports on the PIO-12 default to a condition that forces all of the ports to become input ports. Thus, through default, the PIO-12 I/O board is placed in mode 0, and ports A, B, and C are automatically configured as 8-bit input ports.

I/O WIRING CONNECTIONS

Wiring connections to the MetraByte PIO-12 are provided through a 37-pin, D type connector that is part of the I/O board. Typical of most plug-in boards, the 37-pin connector provides the I/O connection at the rear of the computer. The connector on the PIO-12 board is a 37-pin male. Mating female connectors and cables are available from a variety of sources including MetraByte Corporation.

Figure 5.9 shows the *connector pin assignments* for the 37-pin D type connector that is part of the PIO-12 board. Notice that 24 of the 37 pins are assigned as I/O bits to ports A through C and that the remainder of the pins are power supply connections or serve as a common ground for the digital I/O signals. This pin-out diagram is an essential reference when connecting real-world sensors or loads to the I/O board. MetraByte manufactures a handy accessory identified as the STA-U *screw terminal box*. This accessory, shown in Figure 5.10, serves as a wiring termination board and consists of a printed circuit board with screw terminal strips housed in a box.

The box connects to the D type connector on the I/O board by a ribbon cable. The accessory box provides screw connections for all 24-bits assigned to the three I/O ports along with the power supply connections and digital grounds. This interfacing accessory is invaluable when making wiring connections from sensors or external loads to the computer.

I/O SOFTWARE INSTRUCTIONS

Most every computer language offers an *input instruction* that enables the computer (microprocessor) to receive data from sensors or switches via an interface board. The language also provides an *output instruction* that allows the computer to output data through an interface board to the external world connected to the computer. Since this text focuses on applying the languages C and BASIC to I/O applications, only the input and output instructions provided in these two languages will be illustrated in programming examples. Note that not only does the syntax of I/O instructions vary among languages (Pascal, BASIC, C, and FORTRAN) but the instructions vary among computers. This variation in instructions is a function of which microprocessor is used in a particular computer. Computers that use Motorola processors such as the Apple Macintosh use PEEK (input) and POKE (output) I/O statements while IBM computers and IBM clones use INP (input) and OUT (output) statements.

Therefore, programs written in BASIC for execution on computers that use the well-known Intel microprocessors including the 8086, 80286, 80386, or 80486, would use INP and OUT statements, while programs written in BASIC for computers using Motorola processors, such as the 68000, would use PEEK and POKE statements for I/O.

The fundamental form of the Intel input instruction (INP) in both the C language and BASIC language is indicated;

```
INP(n)
```

where **n**, is an integer which represents a number or port address of the input port being selected or accessed through the software. The following example illustrates

CONNECTOR PIN ASSIGNMENTS

All digital I/O is through a standard 37-pin D-type male connector that projects through the rear panel of the computer. For soldered connections a standard 37-pin D female connector is the correct mating part, and can be ordered from Keithley MetraByte as part number SFC-37.

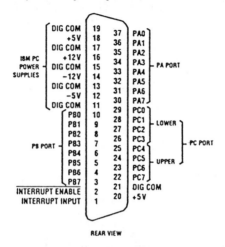

Figure 5.9 Connector Pin Assignments *(Courtesy Keithley MetraByte Corporation)*

the use of the INP statement in both C and BASIC in a more thorough way. Notice that the statement contains a variable name assigned to the data and that data has been input to the computer via the addressed port. The input port address is provided in Hex. This Hex address is consistent with the address ranges described earlier in this chapter.

C Version	*BASIC Version*
`Indata = Inp(0×300);`	`Indata = INP(&H300)`

The **0x** and the ampersand (**&**) used in this statement show that the port address within the parenthesis is provided in Hex format. This instruction also can be written so that the input port address is provided as a decimal number. In the following input statement, the port address within the parentheses does not have the **0x** or the ampersand (**&**); therefore, the computer expects this address to be the decimal equivalent of the port address.

C Version	*BASIC Version*
`Indata = Inp(768);`	`Indata = INP(768)`

CABLE DESCRIPTIONS

C-1800

The C-1800 is an 18 inch ribbon cable with two 37-pin female D-type connectors. This cable is available in a shielded version as S-1800.

S-1800

The S-1800 is an 18 inch shielded ribbon cable with two 37-pin female D-type connectors. This cable is a shielded version of the C-1800.

C-1800

FEATURES

- Provides convenient connection to Keithley MetraByte data acquisition boards
- Complete with plastic enclosure
- Screw terminals accept wire sizes 12-22 AWG
- User breadboard area for custom circuits
- Connects directly to data acquisition board with cables
- Mounts in RMT-02 enclosure

FUNCTIONAL DESCRIPTION

A variety of Screw Terminal Accessories are available for use with our data acquisition boards. These accessories are used to provide access to all signals on the data acquisition board. A STA-Series accessory can also be used in conjunction with a signal conditioning or expansion accessory, such as the EXP-16, to provide access to those signals not available on the signal conditioning accessory.

The STA-Series includes 19 different screw terminal accessories and 2 BNC connector accessories. Each screw terminal accessory varies due to the number of pins on the associated data acquisition board's connector as well as the designation of those pins. Screw terminal accessories that have been designed exclusively for a particular board include labeled terminals and ground connections between analog channels. The two BNC connector boxes are for use with the very high speed DAS-50 and DAS-58 boards.

All STA-Series and BNC products include a plastic enclosure with a cover to protect the wiring. The enclosure can be mounted in a rack (RMT-02) or placed on a flat surface. Additionally, the board can be mounted in a system without the enclosure. All wires can be brought out through cutouts in the side of the boxes.

The following table indicates the compatibility between data acquisition boards and the Screw Terminal Accessories.

Figure 5.10 Screw Terminal Box and I/O Cable *(Courtesy Keithley MetraByte Corporation)*

Executing either of these instructions will cause one byte of binary data (eight bits) to be inputted into the computer through port 300 Hex (768 decimal). The binary data will reside in the microprocessor and is available for manipulation, logical comparisons, mathematic operations, or outputting through an output port. The following version of the identical BASIC INP statement includes the percent sign (%) to indicate that the data assigned to the variable name **indata** is an integer. This statement is not available in the C language.

```
Indata% = INP(&H300)
```

An integer is a number that ranges from −32768 to +32767. Specifying that the variable is an integer saves memory space and causes the program to execute slightly faster. Failure to indicate the percent sign will cause the computer to default to a condition, which uses more memory and results in the program running slightly slower. Programs can be written with or without the variable being declared as an integer. If program execution speed is a serious concern, the program should be compiled rather than interpreted.

Output from the computer through an I/O port is initiated by the *out statement*. The fundamental form of the out statement in both C and BASIC is indicated:

C Version	*BASIC Version*
`outp (port,byte);`	`OUT port,byte`

In these software statements (port) is the output port address in Hex or decimal form and (byte) is the data in Hex, decimal, or binary form that is going to be output though the specified port. Recall that the data byte output to a port *must* be eight bits. Even if not all of the eight bits are used in the output circuitry, the data byte sent to the port must be a full byte. Typical output statements in C and BASIC are shown as follows:

C Version	*BASIC Version*
`outp(0×301,0×FF);`	`OUT &H301,&HFF`

In these statements, the output port address on the I/O board is 301 Hex and the digital data delivered to the port consists of eight ones (11111111). These output statement examples can be rewritten to use decimal numbers and would appear as shown below.

C Version	*BASIC Version*
`outp(269,255);`	`OUT 269,255`

Regardless of whether Hex numbers or decimal numbers are used, the port address must be identified in the software instruction and the data delivered to the port must be known. If it is more convenient, variable names may be assigned to the data byte. The data byte may be identified as a variable:

C Version	*BASIC Version*
`leds = 0×F0`	`LET LEDS = &HF0`
`outp(0×301,leds);`	`OUT &H301,LEDS`

These two statements, in both languages, will cause a binary (11110000) to be delivered to output port 301 Hex. Assuming that LEDs (Light Emitting Diodes) were connected to this port, four LEDs would be lit and four LEDs would be off. The LEDs would remain in this ON or OFF configuration until a new bit pattern is written to the port. Latching of the output data bits is an important role played by the 8255 I/O chip.

I/O BOARD INITIALIZATION

Prior to using INP or OUT instructions, the I/O board must be initialized by writing a data byte to the control register in the Intel 8255 chip. Refer to Figure 5.5 in this chapter for a visual orientation to the internal organization of the 8255 chip. The control register is an 8-bit register that defines the input or output function of three I/O ports identified as (A, B, or C). The address of the control register is always the base address of the board + 3.

Observe that the diagram shows that ports A through C are capable of serving as either input or output ports while the control register is only capable of being written to and serves as a write-only register. Figure 5.11 shows the 8255 PPI address map for this interfacing chip.

The control byte consists of eight bits that must be encoded in Hex or decimal and then written to the control register. Each bit has a role to play as it determines whether a port will serve as input, output, and the mode of the ports. Figure 5.12 provides a reference for determining the binary value of the control byte.

Ports A and B can be set to serve as input ports by writing a binary 1 to the appropriate bits. Or they both can be configured as outputs by writing a binary 0 to the appropriate bits. Port C is divided into two halves, identified as the lower nibble (C0-3) and the upper nibble (C4-7). This port serves a handshaking role. It can be split in half providing two 4-bit handshaking nibbles. By selecting the correct mode the two 4-bit nibbles can assume a variety of handshaking roles.

The following example illustrates the writing of a byte to the control register. This byte places the chip in mode 0 and establishes all three ports in the 8255 as Input ports.

C Version	*BASIC Version*
`outp(0×303,0×9B);`	`OUT &H303,&H9B`

Base Address + Offset	Port Designation	Function
Base + 0	Port A	Read or Write
Base + 1	Port B	Read or Write
Base + 2	Port C	Read or Write
Base + 3	Control	Write Only

(a)

8255A BASIC OPERATION

A₁	A₀	$\overline{RD}$	$\overline{WR}$	$\overline{CS}$	INPUT OPERATION (READ)
0	0	0	1	0	PORT A → DATA BUS
0	1	0	1	0	PORT B → DATA BUS
1	0	0	1	0	PORT C → DATA BUS
					OUTPUT OPERATION (WRITE)
0	0	1	0	0	DATA BUS → PORT A
0	1	1	0	0	DATA BUS → PORT B
1	0	1	0	0	DATA BUS → PORT C
1	1	1	0	0	DATA BUS → CONTROL
					DISABLE FUNCTION
X	X	X	X	1	DATA BUS → 3-STATE
1	1	0	1	0	ILLEGAL CONDITION
X	X	1	1	0	DATA BUS → 3-STATE

(b)

Figure 5.11 8255 PPI Address Map *(Courtesy Intel Corporation)*

All three ports can be configured as output ports by writing zero to the required bits. The following command will cause the three ports (A through C) to be configured for digital output.

C Version	*BASIC Version*
outp(0×303,0×80);	OUT &H303,&H80

The following example is provided as an illustration of how port C can be split into two 4-bit ports. Each of these two half-ports can be configured as a input port or as an output port.

C Version	*BASIC Version*
outp(0×303,0×91);	OUT &H303,&H91

The above instruction will create the following I/O configuration among the three ports:

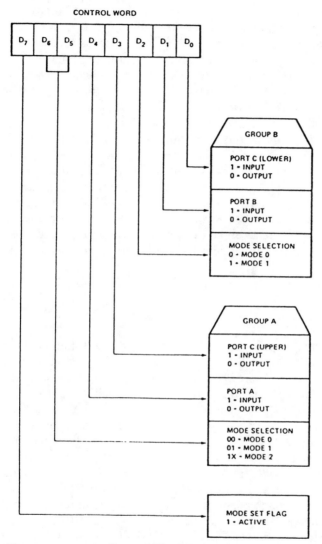

Figure 5.12 8255 Control Word Derivation *(Courtesy Intel Corporation)*

Port A = Input;
Port B = Output;
Port C0-3 = Input; and
Port C4-7 = Output.

In this example Ports A and B are treated as "full" 8-bit ports while port C is divided in two equal 4-bit nibbles. The lower nibble, bits 0-3, is configured as four input bits and the upper nibble, bits 4-7, is configured as four output bits. Note that D7 must be set high (logic 1) to set the configuration of the ports. In an effort to "tie together" many of the points covered in this chapter the brief programs shown in Figure 5.13 and the hardware example shown in Figure 5.14 illustrate the necessary relationship between the software and hardware.

The program illustrated in Figure 5.13 includes the software command needed to initialize the I/O board, input digital data through Port A, and write the inputted data on the computer's CRT screen. The switches shown in Figure 5.14 are typical of digital input devices to an I/O board. The SPST switches shown in the figure can be replaced by limit switches, pushbutton switches, or a variety of other digital sensors. An important consideration included in Figure 5.14 is that the supply voltage for the digital switch circuit is +5 volts. If a direct connection between the external switches and the computer input port is wired, it is imperative that the electrical signal coming into the input port be compatible with the computer.

INPUT HANDSHAKING

Input handshaking is a powerful tool that combines the traditional input of eight bits of data via ports A or B along with the *strobe function*. The input strobe function provides a "hardware controlled" latching of input data rather than total software control of the input process. The strobe function is invoked by placing the 8255 chip in mode 1. In this mode, either port A or port B can be configured for *strobed input*. Figure 5.15 shows a simplified block diagram in which port B serves as an input port, bit D2 on port C (PC2) is the strobe input, and bit D1 on port C (PC1) is the input buffer full bit.

The *input buffer full* (IBF) *bit* will be set high (logic 1) after the data has been latched in response to the strobe input. Observe that in order to use the strobed input, handshaking a hard wire connection is required. In addation to the eight bits of incoming data that must be connected to port B, one bit identified as the strobed input must be connected to bit D2 of port C. A logical low on the strobe input causes the data on port B to be input through the port and latched in the port. Strobed input handshaking is not limited to port B; it can also be accomplished through port A with appropriate wiring and software changes.

The C language program and the equivalent BASIC program shown in Figure 5.16 provide the essential instructions needed to monitor the IBF bit and, when appropriate, cause the computer to input new data via port B. Each time new data is presented at port B the computer responds by reading the data and clearing the IBF *flag*. Resetting this flag bit prepares the IBF bit for the next *data read cycle*.

```
/*                                                              */
/*      Program to establish I/O and Screen Print
        PIO-12 I/O Board Base Address = 300 Hex                 */


#include <stdio.h>
#include <conio.h>

main()
  {
    short int data;

    clrscr();
    outp(0x303,0x90);              /* Initialize ports
                                         Mode 0
                                         Port A input
                                         Port B and C output    */
    do                             /* Start of do loop          */
      {
        data = inp(0x300);         /* Get the data              */
        gotoxy(10,6);              /* Set the screen position   */
        printf("Input decimal value is %d",data);
      }
    while(1==1);                   /* Loop forever              */

  }




'
'
'                 Example 5-13
'
'

REM Program to establish I/O and Screen Print
REM PIO-12 I/O Board Base Address = 300 Hex

CLS                               ' Clear the screen
OUT &H303, &H90                   ' Initialize ports
                                  '     Mode 0
                                  '     Port A input
                                  '     Port B and C output
DO                                ' Start of do loop
  D% = INP(&H300)                 ' Read the input port
  LOCATE 6, 10                    ' Set the screen postion
  PRINT "Input decimal value is "; D%    ' Print the value
LOOP WHILE 1 = 1                  ' Loop forever
END                               ' Program end
```

Figure 5.13 Input Program Example in C and BASIC

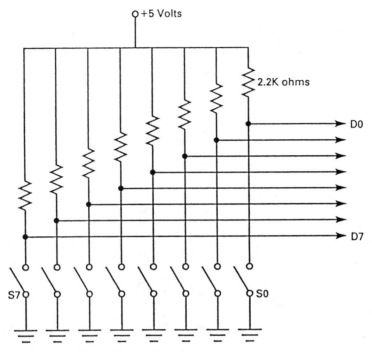

Figure 5.14 Digital Input Switches

BIT ISOLATION

Many data acquisition and control applications do not require the use of all eight bits of input data. While it is not possible to input less than a byte of data, the incoming data can be manipulated so that particular unwanted bits are *masked out*. The masking of unwanted data bits is accomplished by ANDing the incoming data with an 8-bit mask consisting of 0s and 1s. Typically, the masking process is applied to situations where some bits of an input port are not used, floating, or are otherwise not relevant.

The following example shows an incoming data byte coming from a group of switches. The incoming data from the switches is ANDed with a mask consisting of seven binary 0s and one binary 1. The intent of this mask is to block out bits D1 through D7 and to retain only D0 as a valid bit. The ANDing process blocks the unwanted bits and allows the binary value of bit D0 to be retained. This bit reflects the status of the single bit on the DIP switch or other real-world device.

```
Incoming data    10101101
AND
Mask             00000001
Result           00000001
```

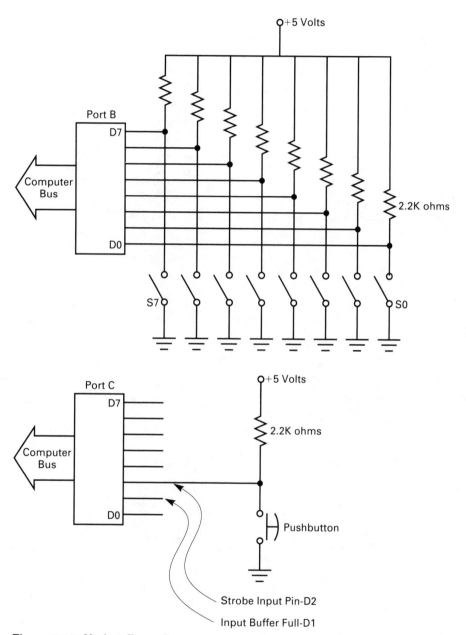

Figure 5.15 Mode 1 Example

```
/*                                                              */

#include <stdio.h>
#include <conio.h>

void display(void);

main()
  {
    int x,z;

    clrscr();
    outp(0x303,0x86);              /* Initialize ports
                                            Mode 1
                                            Port B input
                                            Port A and C output    */
    do                             /* Start of do loop             */
      {
        gotoxy(12,10);             /* Set the screen position      */
        printf("Press any key to terminate Program");
        x = inp(0x302);            /* Get the handshaking info      */
        z = x & 2;                 /* Isolate the IBF bit           */
        if (z == 2)                /* If data is ready...           */
          display();               /*     call function             */
      }
    while(!kbhit());               /* Loop until keypressed         */

  }


void display(void)
  {
    int y;

    y = inp(0x301);                /* Get the input data            */
    gotoxy(12,8);                  /* Set the screen position       */
    printf("The incoming data is %4d",y);
  }
```

(a)

```
DECLARE SUB DISPLAY ()
'
'
'
'
'

CLS                                ' Clear the screen
OUT &H303, &H86                    ' Initialize ports
                                   '    Mode 1
                                   '    Port B input
                                   '    Port A and C output
DO                                 ' Start of do loop
  LOCATE 12, 10                    ' Set the screen postion
  PRINT "Press any key to terminate program"      ' Print message
  X = INP(&H302)                   ' Get handshaking info
  Z = X AND 2                      ' Isolate IBF
  IF Z = 2 THEN CALL DISPLAY       ' If new value ready call subroutine
LOOP WHILE INKEY$ = ""             ' Loop until keypress
END                                ' Program end

SUB DISPLAY                        ' Start of subroutine
  Y = INP(&H301)                   ' Get the input data
  LOCATE 10, 10                    ' Set screen position
  PRINT "The incoming data is "; Y; "  "' Print the value
END SUB                            'End of subroutine
```

(b)

Figure 5.16 Strobed Input Program in C and BASIC

If the status of bit D0 were 0, then the resulting byte would be eight 0s. On the other hand, if bit D0 were a logical 1 then the resulting byte would be 00000001, 01 in Hex, and 1 in decimal form.

The masking process allows the programmer to isolate any incomming bit and to monitor its status while disregarding all other bits. If the incoming value of the bit being monitored is a logical 1, and it is ANDed with a logic 1 the result will equal a logical 1. However, if the incoming value is a zero the result will be a zero. Observe that the masking process does not change the incoming data. True to its name, the masking process only masks out the unwanted bits.

The logical OR function also can be used in conjuction with the masking of digital bits. The OR function "forces" individual bits high when unknown bits are ORed with logical 1s. The following example illustrates logical ORing.

```
Incoming data    10101101
OR
Mask             11110000
Result           11111101
```

In this example, the upper four bits D7 through D4 are forced high by ORing the mask with the incomming data. Through the application of logic functions and through I/O instructions a powerful range of software commands are provided for data acquisition and control applications.

The masking process, using the AND software command or the OR command, provides the programmer with a group of powerful software commands that allow determination of the open or closed status of switches and other digital on/off devices.

INTERFACING PROBLEMS

Problem #1: Wire the necessary circuitry so that eight DIP switches will serve as input sensors connected to port C. Write a program that will input data from port C and display the decimal value of the data on the computer screen. Provide instructions in the program so that if the incoming data exceeds the value of 128 decimal, the following message should appear on the computer screen:

```
The data value is greater than 128
```

Design the program so that when the incoming data falls below 128 the message will be removed from the screen. The suggested screen appearance is:

```
Data: 131
The data value is greater than 128
```

Problem #2: Wire the necessary circuit so that eight DIP switches are connected as inputs to port A and a second group of eight DIP switches are connected as inputs to port B.

Write the necessary program so that the incoming data on port A is ADDED with the data on port B. Display the result in decimal form on the computer screen. If the sum of the two data bytes exceeds the value of 128 decimal, print the following message on the screen:

```
Data Sum Exceeds 128
```

Design the program so that when the data sum falls below 128 the data sum message will be removed from the screen. The suggested screen appearance is:

```
Sum: 131
Data Sum Exceeds 128
```

Problem #3: Modify the above program (Problem #2) so that the two incoming data bytes are checked for equality. If the two bytes are equal in value, the following message should be printed in the computer screen:

```
Data Bytes are Equal
```

If the incoming data bytes are not equal, the program should decide which port has the greater value and display the appropriate message on the screen. If the data on port A is larger than the data on port B the message should be:

```
Data on Port A is Larger
```

If the data on port B is larger the message should read:

```
Data on Port B is Larger
```

Design the software so that the program is in a loop; checking the incoming data, and constantly updating the screen message.

Problem #4: Wire the necessary circuit so that port B is connected to eight DIP switches and a N.O. SPST switch is wired as a source of strobe pulse.

Write the necessary program so that each time the N.O. push button is depressed the binary data on the DIP switches will be read and the decimal value displayed on the computer screen.

Problem #5: Modify the circuit and program of Problem #4 so that data is input via port A rather than port B.

CHAPTER 6

Interfacing to the PC Bus: Digital Output

The outputting of digital TTL compatible binary bits from the computer to the "real world" can be a powerful tool, beyond the traditional roles that computers play in word processing, accounting, and other business applications. Output ports on data acquisition and control I/O boards handle output data bits in the same way that input ports handle TTL compatible input data. The binary bits available at an output port are often latched by an 8255 chip which was described in Chapter 5. The output provided by the output port can be in the form of TTL level binary bits or they can be in the form of relay contacts. Relay outputs are available in two different formats. One form of relay output uses an electromechanical relay for each of the eight binary bits output by the port. Another form of relay output is the solid-state relay. In this form, the output port consists of eight solid-state relays, one for each bit of the output byte.

If the output from the I/O port is provided as a traditional TTL signal, then this signal is defined in terms of standard TTL units. When viewed in terms of current and voltage level, this specification provides a current sinking limitation of 1.7 mA while providing a logic 0 output. A sourcing limitation of 200 uA is required while maintaining the logic 1 TTL voltage threshold of $\geq$ 2.4 volts.

Figure 6.1 illustrates an output port with LEDs connected as loads. The LEDs are buffered by an invertor and are lit by a TTL logic level 1 provided as an output from the port. Other electrical loads may be connected in place of the LEDs if proper buffering or signal conditioning is provided. For those applications requiring an increased fan-out from the output port a buffered I/O board is required.

An I/O board with increased current sinking capabilities is the CYDIO 24H produced by CyberResearch, Inc., 25 Business Park Drive, Branford, CT 06405. This board uses discrete logic instead of the Intel 8255 integrated circuit interface chip to provide the interfacing and latching of output binary data. Using discrete logic rather than an integrated circuit allows each output bit to sink up to 64 mA of current while maintaining TTL compatibility. This extended current sinking rating allows this board to effectively "drive" a variety of indicator lamps, electromechanical relays, as well as other devices within this current rating and TTL voltage rating. Full software compatibility is maintained between this board and other I/O boards that use the Intel 8255 chip. Compatibility between boards from different manufacturers simplifies system design and software development.

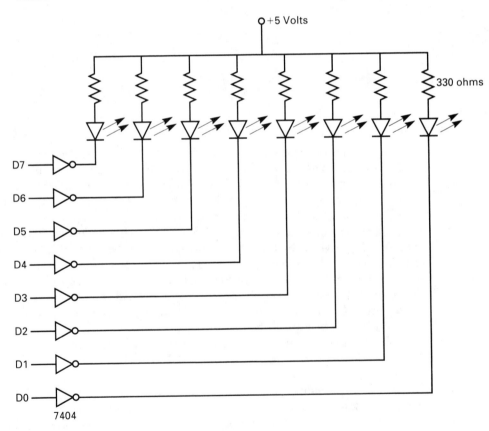

Figure 6.1 TTL Output Port with Buffers

Description—CIO-DIO24H

The CIO-DIO24H is for applications that require higher currents than are available from a simple 8255; currents up to 60 mA. The CIO-DIO24H is able to light LED's, switch relays and accomplish a host of tasks beyond the capability of an 8255 based I/O board.

The CIO-DIO24H is 100% software compatible with the 8255 mode 0 and so is supported by Labtech and other programs.

Specifications

High Drive: I/O—CIO-DIO24H	
Output low voltage	0.55v
Output low sink current	64 mA
Output high voltage	2.0v
Output high source current	15 mA
Interrupt inputs	1, PC Bus 2-7
Power Consumption +5v	400 mA
Compatibility	PIO-12, PIO-24

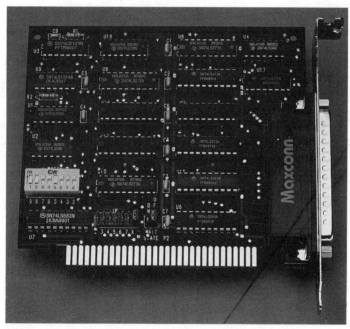

Figure 6.2 CyberResearch CYDIO 24H Digital I/O Board
(Courtesy CyberResearch, Inc.)

Chapter 7 will explore I/O buffering considerations in more detail and will include techniques that are available for interfacing to a variety of real-world devices.

DEFINING OUTPUT PORTS

Depending upon the selection of the mode of the 8255 chip, and the bit pattern written to the control register in the chip, any or all of the ports on the typical I/O board can be configured as output ports. The following instruction is provided as an example of the technique used to set two of the ports on the I/O board to an output configuration. This example is based on the assumption that the I/O board being configured has a base address of 300 Hex. In this example, ports A and B are configured as output ports and port C is configured as an input port.

C Version	*BASIC Version*
`outp(0×303,0×89);`	`OUT &H303,&H89`

For a review of the technique used to select the mode of the 8255 PPI chip and the bit pattern used to select the I/O directions of the three ports, refer to Chapter 5.

The data byte output through a port by an OUT instruction can be imbedded in the OUT command. This bit pattern can be written in either Hex or decimal form. As illustrated in programming examples in Chapter 5, if the data value is in Hex, the syntax must include the ampersand Hex (&H). Without this coding, the data byte will default to a decimal value. The maximum value of a data byte that can be output through a single port is Hex FF. Stated another way, the maximum value for any byte of data is 255 decimal, FF Hex, or 11111111 in binary. The following programming example is offered as an illustration of how a data byte is imbedded in an OUT instruction. Referring to the Figure 6.1, assume that the intent is to light every other LED connected to the output port. This can be accomplished by the following program which is shown in both C and BASIC.

C Version	*BASIC Version*
`outp(0×303,0×80);`	`OUT &H303,&H80`
`outp(0×300,0×AA);`	`OUT &H300,&HAA`

This program needs to be executed only once as the bits are latched by the circuitry of the 8255 chip. The status of the output bits will remain at either a logic high or logic low until the computer is reset, turned off, or a new data byte is written to the port.

OUTPUT HANDSHAKING

Handshaking plays the same powerful role when data is output from a computer as it did during the inputting of data that was described in Chapter 5. When handshaking is used in conjunction with the output processes, ports A and B serve their regular role as output ports while port C provides the handshaking function.

When port C is placed in mode 1 and configured for output handshaking, a change occurs in the function of the handshaking bits in this port. The bits in port C that were used as STB and IBF during the input process (see Chapter 5) now serve new roles and are identified as OBF and ACK.

The OBF (Output Buffer Full) bit goes low to show that the computer has written data out to a specified output port. This bit serves the role of a flag. The flag indicates whether the data has been "delivered" to the output port. This flag can be queried by software to determine if the data has been sent or this single bit "flag" can be connected to external hardware for monitoring purposes.

The ACK (acknowledge) bit serves as a handshaking input bit to the computer. A low on this input bit indicates that the peripheral device is ready to receive the data. An active low on this bit also resets the OBF "flag" from its logic low state to a logic high state. The OBF bit can play a particularly important role in outputting process. Monitoring of this bit through software can allow the program to control when the computer writes data to the peripheral device.

Figure 6.3 shows a simplified drawing of the mode 1 output handshaking bits and their roles. Observe that when Port A is configured for output, bits on Port C are used for output handshaking. In the output handshaking configuration, bit PC7 serves as the output buffer full (OBF) bit and PC6 serves as the acknowledge (ACK) handshaking input to the computer. It should be remembered that the OBF bit is an output bit from the 8255 and the ACK bit is an input bit to the chip.

The example program shown in Figure 6.4 is written in both C and BASIC. This program outputs data from Port A *only* when the peripheral device, assumed to be a N.O. SPST push button switch, generates an active low. The active low pulse is connected to the ACK bit (PC6) of the Intel 8255 PPI chip on the I/O board.

BIDIRECTIONAL HANDSHAKING

In addition to the previously mentioned input or output functions and handshaking capabilities, the I/O ports on products like the Keithley MetraByte PIO-12 board can be configured for bidirectional data flow in conjunction with handshaking control. The handshaking bits OBF, ACK, IBF, and STB previously defined play a role in the bidirectional mode of operation.

Bidirectional I/O is typically defined in the technical literature as mode 2 (*strobed bidirectional bus I/O*). This mode is activated by making bit D6 in the

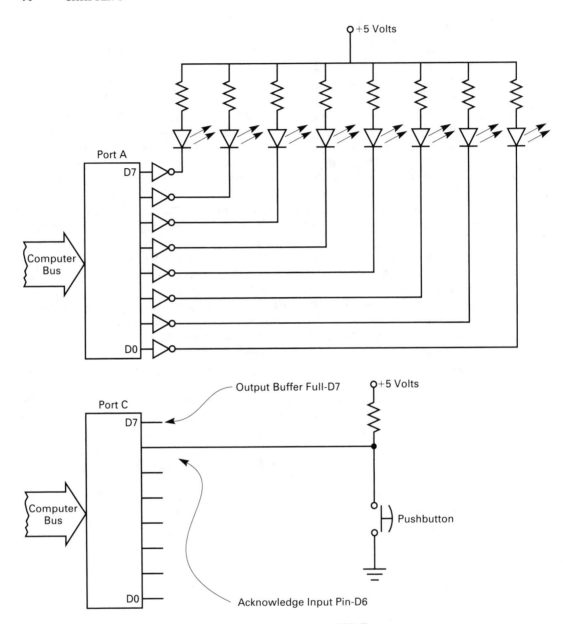

Figure 6.3 Mode 1 Output Handshaking Using the 8255 PPI Chip

control register a logical high. Once the 8255 chip is in mode 2 the determination of whether Port A is serving as an input port or output port resides with the logic level on the ACK (PC6) pin. A logic low on this input enables the tri-state output buffer of port A to send out the binary data. If ACK is not low the buffer will be

```
/*                                                           */
/*      Output Handshaking                                   */

#include <stdio.h>
#include <conio.h>

main()
  {
    int dat, buf;

    outp(0x303,0xAD);               /* Initialize ports
                                          Mode 1
                                          Port A and C output
                                          Port B input         */
    do                              /* Start of do loop        */
      {
        clrscr();                   /* Clear the screen        */
        printf("Enter a number to output  "); /* Prompt user   */
        scanf("%d",&dat);           /* Get number              */
        do
          {
            buf=inp(0x302);         /* Port C has status of A's OBF */
            if((buf & 0x80) == 0x80) /* If MSB = 1 then         */
              outp(0x300,dat);      /*          A's OBF was empty*/
            else                    /* else                    */
              {
                gotoxy(12,10);                  /* Set screen position*/
                printf("Buffer is full"); /* and print message  */
              }
          }
        while((buf & 0x80) !=  0x80);
      }
    while(dat>=0);                  /* Loop forever            */

  }
```

                                    (a)

```
'
'
'
'
'

REM Digital Output with Handshaking
REM
REM LED's connected to Port A
REM Momentary switch to PC6

OUT &H303, &HAD                'Set Port A and B output Mode 1
DO
  CLS                          ' Clear the screen
  INPUT "Enter a number for output ", dat  ' Get the number to output
  DO
    buf = INP(&H302)                 ' Port C has status of A's OBF
    IF ((buf AND &H80) = &H80) THEN ' If OBF is empty then ..
      OUT &H300, dat           '          output the data
    ELSE
      LOCATE 12, 10            ' Set screen position
      PRINT "Buffer is full"
    END IF
  LOOP WHILE ((buf AND &H80) = 0)
LOOP WHILE (dat >= 0)
END
```

                                    (b)

Figure 6.4 Output Handshaking Program in C and BASIC

in the high impedance state. The following control byte will place the 8255 chip into mode 2 and will prepare port A as a bidirectional port.

C Version   *BASIC Version*

outp(0×303,0×C0);   OUT &H303,&HC0

Figure 6.5 shows a simplified block diagram of Port A in a bidirectional mode and the four handshaking control pins that are part of Port C. Observe that acknowledge (ACK) and strobe (STB) are both real inputs to the port. In this context, the word "real" means that the STB and ACK pins on the 8255 must be connected to external switches or electrical signals created by circuitry external to the computer. The following explanations further describe the role that the OBF, ACK, STB, and IBF handshaking bits play as the bidirectional port is used to carry data in either direction.

Manipulation of the ACK pin and the STB pin controls the direction of data flow through the bidirectional port. Because both of these input pins are active low, activated by a logical 0, both the ACK and STB can not be low at the same time. The bidirectional port will only pass data in one direction at a time.

INTERFACING PROBLEMS

Problem #1. Wire the necessary circuitry so that eight LEDs will serve as output devices connected to port C.

Write a program that will output data to the LEDs. The data written to the LEDs should be such that every other LED is lit. Write the program so that the program halts after one execution of the program. Demonstrate the circuit to the instructor!

Problem #2. Modify the program written for Problem #1 so that it will alternately blink the eight LEDs. The first LED pattern displayed should have the odd numbered LEDs (D1, D3, D5, D7) lit. The second pattern should have the even numbered LEDs (D0, D2, D4, D6) lit. Include a loop in the program so that the blinking will continue until any key on the keyboard is pressed. Design a timing loop in the program so that the blinking is reasonably slow and observable to the human eye. Demonstrate the circuit to the instructor!

Problem #3. Wire the necessary circuitry so that eight DIP switches are connected to an input port and eight LEDs are connected to an output port. Write the necessary program so that the data input from the DIP switches is displayed on the CRT as a decimal number. The data should simultaneously be displayed in binary form by the LED output port. The program should include a loop so the each time the incoming data from the DIP switch is changed both the numerical value on the screen and the binary value indicated by the LEDs is updated. Demonstrate the circuit to the instructor!

Outputting Data

OBF (Output Buffer Full). The OBF will go low to indicate that the microprocessor has written data out to Port A.

ACK (Acknowledge). A low on this input enables the tri-state output buffer of Port A to send out the data. Otherwise, the output buffer will revert to the high impedance state.

Inputting Data

STB (Strobe Input). A low on this input loads data into the input latch.

IBF (Input Buffer Full). A high on this output pin indicates that the data has been loaded into the input latch.

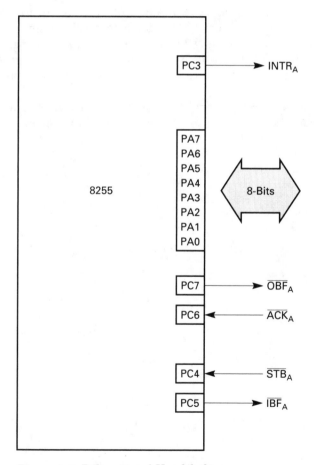

Figure 6.5 Bidirectional Handshaking

Problem #4. Wire the necessary circuit so that port B of the I/O board is connected to eight LEDs and a N.O. SPST switch is wired as a source of ACK pulses for handshaking purposes. Write the necessary program so that each time the N.O. pushbutton is depressed the computer will output progressively higher binary data to the LEDs. The LEDs should emulate the following sequential binary bit pattern in response to "strobbing" by the ACK pin.

LEDs   00000001 First Depression of Pushbutton
00000010 Second Depression of Pushbutton
00000011 Third Depression of Pushbutton
00000100 Fourth Depression of Pushbutton
etc.......................................

Demonstrate the circuit to the instructor!

Problem #5. Wire the necessary circuit and write a program to implement bidirectional data flow through port A. This assignment will require the inclusion of two N.O. SPST switches, use of eight LEDs, and the use of an eight bit DIP switch. The program should be written so that data can be input to the computer and displayed on the CRT or data can be output by the bidirectional port to the external LEDs.

This is not a trivial assignment. To construct this circuit you will need a tri-state buffer, an 8-bit latch, and some thought applied to the circuit design. Demonstrate the circuit to the instructor!

CHAPTER 7

Digital Signal Conditioning

The personal computer, when coupled with a data acquisition and control board, forms the basis of a powerful tool which has application in all aspects of industrial control, manufacturing automation, and other areas of science and technology. The electrical output voltage and current rating for most DA & C boards is limited to modest electrical ratings. Figure 5.1 showed a popular and often cloned digital I/O board, the PIO-12 manufactured by Keithley MetraByte Corporation. The output electrical ratings for this board are 1.7 mA (sinking) while maintaining ≤ .45 volts, and 200 uA (sourcing) while holding the output voltage ≥ 2.4 volts. These specifications are TTL compatible and are within the commonly accepted TTL specifications.

Figure 7.1 provides the electrical output specifications for a typical DA & C board. Observe that the voltage output high is + 2.4 volts or greater while sourcing 400 uA of current. Correspondingly, the voltage output low is + .45 volts or less while sinking 1.7 mA of current. The output from this board, like the output from individual TTL chips, has the capacity to sink external loads much better than source loads. The same circuit design techniques used with individual TTL chips should be used with output from this board.

SPECIFICATIONS

Logic Inputs and Outputs	Min	Max
Input logic low voltage	-0.5V	0.8V
Input logic high voltage	2.0V	5.0V
Input load current PA, PB, PC port		
(0 < V in < 5 V)	$-10\ \mu$A	$+10\ \mu$A
Input low current, interrupt inputs	—	-0.4 mA
Input high current, interrupt inputs	—	20 μA
Output low voltage PA, PB, PC ports		
(Isink = 1.7 mA)	—	0.45 V
Output high voltage PA, PB, PC ports		
(Isource = 200 μA)	2.4 V	

All outputs and inputs are TTL/DTL-compatible and outputs will drive one standard TTL load (74 series) or 4 LSTTL (74LS) loads. CMOS compatibility can be obtained by connecting a 10 k ohm pull-up resistor from the input or output to +5 V.

Outputs do not have the drive current to control most solid state relays directly (e.g., OACs, ODCs). Interface boards such as the SRA-01 and SSIO-24 provide buffers which allow the PIO-12 to control these relays.

Figure 7.1 PIO-12 TTL Logic Level Characteristics *(Courtesy Keithley MetraByte Corporation)*

OUTPUT SIGNAL CONDITIONING

The previously stated current and voltage values are satisfactory for interfacing to some devices, particularly TTL compatible devices such as integrated circuit chips. However, these voltage and current specifications are totally unsatisfactory when attempting to interface the DA & C board to real world equipment such as electric motors connected to pumps, fans, and other rotary devices, electromechanical solenoids, and other heavy-duty electromechanical devices. The current and voltage rating of TTL compatible outputs simply can not drive these devices.

ELECTROMECHANICAL RELAYS

For decades electromechanical relays have been the primary interface between electrical circuits, especially solid-state circuits, and heavy duty real world devices such as motors, pumps, and solenoids. Electrical interfaces between the TTL world of the computer and the real world of electromechanical relays are possible if proper buffering and current drivers are used. Figure 7.2 illustrates a single bit from a TTL compatible output port and two suggested solutions that are available

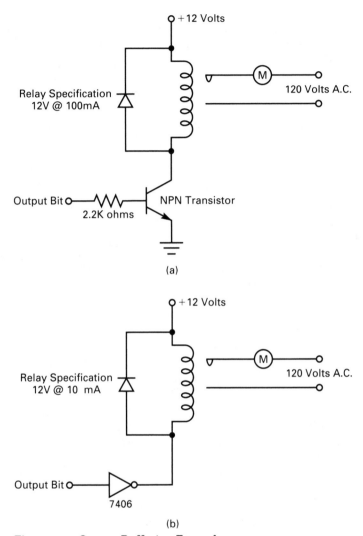

+12 Volts

Relay Specification
12V @ 100mA

120 Volts A.C.

Output Bit
2.2K ohms

NPN Transistor

(a)

+12 Volts

Relay Specification
12V @ 10 mA

120 Volts A.C.

Output Bit

7406

(b)

Figure 7.2 Output Buffering Examples

for interfacing between this bit and a 12 volt DC relay. Figure 7.2(a) uses a transistor driver between the TTL output bit on the I/O port and the coil of the electro-mechanical relay. The relay specifications of +12 volts DC at 100 mA is typical of a number of electromechanical relays. In the example of Figure 7.2(a), the NPN transistor must provide the "drive" current for the relay while handling +12 volt potential when the relay is not active. The two major specifications required for selecting the proper transistor for an application like this are an Ic max of ≥ 120 mA and a Vcbo of ≥ 24 volts. These two specifications include a generous margin of safety and will be adequate for the specified relay. Other relay coil voltage and current ratings can be chosen in place of the 12 volt example shown in this figure. If a relay is chosen that has a higher current and/or voltage rating, then a transistor must be selected that has a current and voltage rating compatible with the relay.

Part (b) of Figure 7.2 illustrates the use of an open collector TTL integrated circuit as the interface between the computer and the relay. Here the suggested chip is a 7406 inverting buffer/driver. This chip has the unique ability to drive DC loads that require operating voltages up to +30 volts. The 7406 also can sink currents up to 30 mA. The relay used in Figure 7.2(b) has a coil specification of +12 volts DC at 10 mA. The 7406 chip can easily handle this interface task and would be an ideal interface between the PIO-12 I/O board and the specified electromechanical relay. Even though the 7406 chip is a TTL chip, it has the ability to provide the "drive" current of 10 mA and to handle the voltage of +12 volts that are needed to operate the relay.

In both circuits shown in Figure 7.2, the relay is bypassed by a back-biased diode. This diode typically is included in an electromechanical circuit like this for the purpose of shunting the back EMF current, caused by the collapsing electromagnetic field, around the relay coil. Including the diode in the circuit is an important precaution and it should be used whenever possible!

DA & C boards equipped with electromechanical relays are also available from equipment vendors such as MetraByte Corporation, CyberResearch, Inc., Industrial Computer Source, and Computer Boards, Inc. Figure 7.3 shows the CIO-PDISO8 produced by Computer Boards, Inc.

This relay-equipped board lends itself to easy integration into microcomputer-based control applications. The I/O board shown in Figure 7.3 has eight output relays that have contact ratings of 3 amps at 120 volts AC, or 3 amps at 28 volts DC. This board is also equipped with eight optically isolated input bits. The isolated input bits are rated at 500 volts of isolation and can handle either DC or AC input signals up to 24 volts. The relay contacts on this board can be directly wired to external loads such as motors, solenoids, or other devices that require electrical currents and voltages within the limitations of the relay contacts.

Another form of relay interface, identified as an external relay board, is available to those who need to connect TTL output levels from a computer to the higher current and voltage levels of the real world. Figure 7.4 shows a picture of an external relay interface board produced by Computer Boards, Inc. This product, identified as the CIO-ERB24, has 24 electromechanical relays mounted on a 19″ × 4.5″ printed circuit board.

This product is connected through a ribbon cable to an interface board that is installed in the computer. In Figure 7.4, the relay board is shown together with the ribbon cable and a digital interfacing board. The interfacing board can be one of a variety of boards including the PIO-12 produced by MetraByte Corporation. Observe that a 37-conductor ribbon cable connects the two units together. All 24 relays on the external board are available for interfacing applications. They can

Figure 7.3 Isolated Relay Interfacing Board

be used if all three ports (A, B, and C) on the digital I/O board are configured as output ports.

The CIO-ERB24 board has a high degree of flexibility in that the relay board need not be configured so that all 24 relays on the board are configured as outputs. The CIO-ERB24 in conjunction with the PIO-12 can be configured so that any 8-bit port can be set up as an input port or can be configured as an output port through the relays. Some control applications may require more than 24 relays. This situation can be handled by installing additional PIO-12 boards in the computer along with additional CIO-ERB24 Relay Boards. The flexibility of using 24 relays on the CIO-ERB24 board complete with pre-wired driver circuitry is a desirable option when designing a system that requires control of high current or high voltage equipment. Inclusion of a CIO-ERB24 may be well worth the higher initial equipment cost.

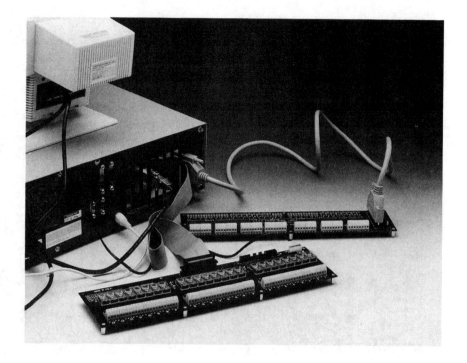

Figure 7.4 External Relay Board

SOLID-STATE RELAYS

Solid-state relays are attractive replacements for traditional electromechanical relays. The SSR (solid-state relay) duplicates the role and mission of the electromechanical relay without moving parts and many of the interfacing disadvantages associated with electromechanical relays. When properly used, the SSR offers desirable features such as long life, reduced electromagnetic interference, lack of contact bounce, and lack of contact arcing. The life expectancy of an SSR is typically 1,000,000 cycles, which under normal use translates into a very long life expectancy. Along with these advantages, solid-state relays do have some disadvantages. For comparison purposes, a listing of a few advantages and disadvantages of solid-state relays are:

Advantages	Disadvantages
Zero voltage turn-on	High internal voltage drop
Long life	May require heatsink
No contacts	Off-state leakage current
Microprocessor compatible	Fairly high cost
Fast response	Only SPST available
No moving parts	Can't switch small signals

Fundamentally similar to a electromechanical relay, the solid-state relay provides electrical isolation between the computer circuitry and the load consisting of a motor or other device. A solid-state relay is composed of four major parts. These parts, shown in Figure 7.5, are the input LED, the optical coupler, the photodetector, and the power handling device, typically an SCR or Triac. Notice in Figure 7.5 that the LED, when forward biased, provides a source of light that is optically coupled to the photodetector.

The active photodetector is connected to the SCR or Triac which controls the flow of current and, therefore, the application of voltage to the electrical load device.

A wide variety of solid-state relays are available. The most popular SSR as an interface device between the digital (TTL) output of a computer and the real world is the AC solid-state relay (SSR) which has a TTL compatible DC input and an AC voltage output. This device, pictured in Figure 7.5, provides a TTL compatible input (typically 3 to 32 volts DC) and a controlled AC output. The AC output typically is rated at 120 or 240 volts, with AC current ratings of 10, 20, 25, or 40 amps.

Observe the four screw connections on the body of the device. These connections are wired as two input terminals and two output terminals. The rectangular block form shown in this figure is the most popular format of the SSR. Other forms are available including modules that mount on printed circuit boards.

Figure 7.6 illustrates the connection between an output bit from a DA & C interfacing board and a solid-state relay. The wiring connection between the output bit and the solid-state relay is simple and requires no special conditioning circuitry. Because of internal current limiting characteristics in the solid-state relay, the output bit from an I/O board similar to the MetraByte PIO-12 can be connected directly to the relay without the need for a current limiting resistor or other interfacing considerations. Current limiting is accomplished internally in the SSR; therefore, it has the ability to adapt to the wide range of input DC voltages. Typically, these voltage range from +3 to +32 volts DC.

In Figure 7.6 the relay is connected to the output bit of the DA & C board in the sinking mode. The sinking mode, often referred to as an active low, is also called negative true-value logic (NTL). This terminology indicates that to operate the solid-state relay, to turn on its internal LED, and to cause the connected AC load to operate, the logic level on the input pin of the SSR must be brought low. Therefore, the output bit from the PIO-12 board must be NTL to turn-on the solid-state relay. This connection arrangement is quite common and agrees with previously described TTL current sinking characteristics.

In the application shown in Figure 7.6, the load is a 1/4 hp, 120 volt AC motor drawing approximately 5.8 amps. This application is ideal for a 120-volt AC solid-state relay that is rated at 10 amps. Besides the ease of application and its inherent flexibility, the solid-state relay provides zero switching. This characteristic means that the SSR switches the AC load current ON or OFF only when the AC sinewave is crossing a zero voltage reference. The implication of zero switching is that

SERIES 1
SCR Output
Solid-State Relays

2.5 Thru 90 Amp
24-480 VAC Output

General Description

Crydom's Series 1 premium line, solid-state power relays incorporate inverse-parallel SCR output devices in the original standard Crydom package with the same highly reliable, noise-immune drive circuitry that has been a Crydom feature for more than a decade. Snubbers are included for high dv/dt applications and inductive loads, with a choice of models offering zero-voltage switching to reduce high inrush currents and electrical noise or Phase Controllable models for random turn-on.

The oversized output chips, together with the Crydom optimized thermal management system, allows a narrower band of temperature excursions, resulting in a significant reduction in thermal cycling fatigue, thereby extending relay life. These premium devices are recommended for use in high temperature, highly inductive load situations where the ultimate in thermal and surge performance is required.

Zero Voltage Models

The inherent zero-current turn-off characteristic of SCRs, and the total absence of arcing mechanical contacts, substantially reduces electro-magnetic interference and back EMF transients. AC input models can be controlled from a wide range of AC Signal Sources (90-280 VAC), and are available in Form A (normally open) configuration only. DC input versions will operate from IC logic signals, and are available in either Form A (SPST-normally open) or Form B (SPST, normally closed) output configurations.

Phase Controllable Models

The -10 versions of DC Input Series 1 Relays are "non-zero voltage" (random) turn-on types, and are optimized for operation from a phase-controlled signal applied at each half of the line cycle. They have been designed for phase control of incandescent lamps or any load with a power factor between 0.6 and 1.0.

Opto-Isolated 4000 VRMS ■
Random and Zero Voltage Switching (AC) ■
Form A and B Output Switching ■
U.L., CSA and VDE Approved ■
Superior Thermal and Surge Ratings ■
AC and DC Wide Control Range ■
400 Hz Relays Available ■

Wiring Diagram

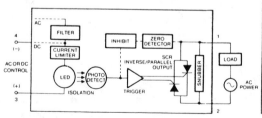

Figure 7.5 A.C. Solid-State Relay (SSR) *(Courtesy Crydom Company)*

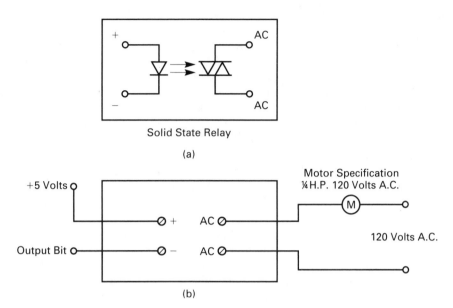

Figure 7.6 I/O Output Bit Connected to SSR

inrush current is eliminated and the resulting harm to mechanical equipment is minimized. Electromagnetic interference that would normally result from high inrush currents is also eliminated.

For those situations where large DC currents need to be controlled by DA & C boards, manufacturers produce DC solid-state relays. This form of solid-state relay can control DC currents. Control of the load current inside the relay is accomplished by using heavy duty transistors inside the SSR rather than triacs. The DC relay does not have a latching function; and, therefore, the TTL logic level must be continuously applied to the input of the SSR. Elimination of the "turn on" logic level to a DC SSR will cause the relay to "disconnect" the DC load. The reader should be aware that the DC relays will not handle AC loads and conversely, AC solid-state relays will not accommodate DC load currents.

INPUT SIGNAL CONDITIONING

Input digital signals need to be conditioned just as output signals need to be conditioned. The need for conditioning of input signals, more often than not, is because the input voltage being applied to the digital ports on a DA & C board is not compatible with the TTL requirements of the board. This lack of electrical compatibility can be solved in a couple of ways.

OPTOISOLATORS

Optoisolators are solid-state devices that can be thought of as the semiconductor version of a transformer. These devices provide the electrical isolation of a transformer, as well as providing a change in voltage level between input and output similar to that found with a transformer circuit. While optoisolators are often used to condition digital signals, they are not capable of handling AC voltages nor are they capable of handling the high current or high voltage ranges normally associated with traditional transformers. However, the optoisolator can provide some of the desirable characteristics of a transformer such as isolation and modest voltage changes. Figure 7.7 shows a typical optoisolator application in which the digital signal created by the mechanical switch is 24 volts DC.

Because of a lack of voltage compatibility this signal needs to be conditioned into compatibility with the TTL requirements of the DA & C board. In this circuit, the optoisolator, consisting of an LED and an optically coupled photo-transistor, provides the role of a "transformer" by providing a technique of conversion of the +24 volt digital signal to a +5 volt pulse. The circuit in Figure 7.7 includes a current-limiting resistor in series with the input side of the optoisolator. This resistor is necessary to limit the current through the internal LED within the optoisolator.

Optoisolators such as the popular model 4N33 or 4N35 do not have internal current limiting circuitry like the solid-state relay previously described. Selection of the proper current limiter resistance value, which is dependent upon the voltage of the input DC signal, is based on the formula shown in Figure 7.7. Optoisolators similar to the 4N33 and 4N35 typically provide a maximum isolation (insulation) value of roughly 1500 volts. Some optoisolators may have isolation values ratings up to 7500 volts. Either of these isolation ratings is sufficient for most situations where a level of protection is need for the computer and data acquisition boards.

Optoisolators find other applications besides input interfacing when different voltage levels are involved. They can be used as a small signal solid-state relay in situations where the load currents are small. In some applications where low level currents and modest voltages are found, the optoisolator may replace the previously described solid-state relays.

MICROCOMPUTER I/O MODULES

Another popular product for interfacing digital signals either in or out of a computer is through digital I/O modules. These modules are produced by several manufacturers. One of these manufactures is Gordos, Inc., 1000 North Second Street, Rogers, AR 72756. The solid-state I/O modules are specifically designed for either input signal conditioning or output signal conditioning. Each type of module is designed to fill an interfacing role between sensors and the digital input to the computer or between the digital output from the computer and the electrical load. Figure 7.8 shows a few of the different types of solid-state I/O modules.

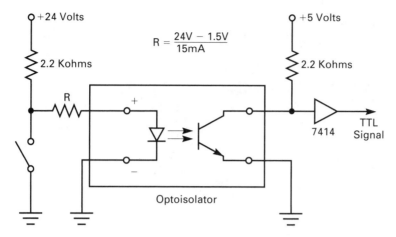

Figure 7.7 Optoisolator Input Signal Conditioning

These modules, as shown in Figure 7.8, are optimized for the TTL requirements of the computer. They combine the isolation qualities of the previously described optoisolators and the power handling capabilities of solid-state relays into a single module. The modules are available in a variety of voltages and types so that almost any input voltage or signal can be accommodated with an "off-theshelf" I/O module. Functionally, the I/O modules fall into four major categories. These categories are:

Input Modules	Output Modules
AC input to computer (TTL)	Computer (TTL) output to AC
DC input to computer (TTL)	Computer (TTL) output to DC

In an effort to ease selection and inventory, the various modules are color coded and labeled according to their function and AC or DC capability. Figure 7.9 describes a typical input module. Note that the modules have pins protruding from the bottom of the module. These pins insert into sockets on a mounting board which is external to the computer.

As with solid-state relays, the inputs to the I/O modules are equipped with internal current limiting circuitry and, where needed, they are equipped with rectifying circuits so that the AC or DC signals can be directly connected to the module. User provided external current limiting resistors are not necessary when using these modules. In a normal installation, the conditioning modules are mounted in an a rack that is external to the computer. Typically, the rack is housed in a cabinet or other suitable enclosure. Wires can then be run from the I/O module assembly to the sensors, indicator lights, motors, solenoids, or other loads. The computer is connected to the conditioning module board by a multi-conductor ribbon cable. Refer to Figure 7.9 and observe that both the AC and DC modules are identified

GORDOS

DIGITAL I/O MODULE MOUNTING BOARDS

FEATURES

- Plug-compatible logic connections on 8, 16, 24 and 32 position boards. Screw terminal barrier block for logic connections on 4-positon boards.
- Screw terminal barrier block for load connections
- Resident pull-up resistors
- 5 amp field-replaceable fuses (LITTLEFUSE #251005 or equivalent)
- LEDs indicate logic status
- All even-numbered logic connections are logic ground

- Input and output modules accepted interchangeably
- Operate with 5, 15 or 24 volt logic supplies
- Captive-screw retaining system for standard-size modules and "Quad-Packs". Pin retaining system for "SM" series miniature modules. Optional hold down bar for "M" and "SM" series miniature modules.
- PB-4, PB-4R, PB-8, PB-16S, PB-16T, PB-24, PB-24Q and PB-32Q UL recognized (E79183) and CSA certified (38595). Additional approvials pending. Consult factory for updated list.

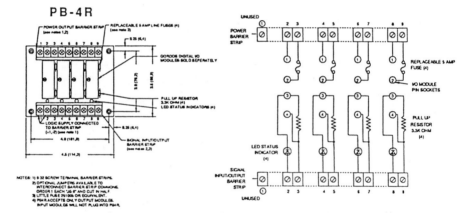

Figure 7.8 Solid-State I/O Modules *(Courtesy Gordos (Crouzet Corporation))*

by different colors. Also note that most every reasonable combination of input voltage and output voltage range is provided by the various modules.

The output I/O module shown in Figure 7.10 indicates some specifications associated with output modules as well as the color of both the DC and AC versions of the module. A variety of sizes and forms of module mounting racks are available. Figure 7.11 shows one such mounting rack.

Input/output racks and modules are popular with circuit designers, systems integrators, and applications engineers because of the "one-stop shopping" provided by the variety of modules and the ease with which modules can be changed. This feature allows for prompt servicing of computer-controlled equipment and simplifies system design.

GORDOS

DIGITAL I/O MODULES

SM SERIES INPUT MODULES

FEATURES

- Plug into Mounting Boards for SM or 0.6" Modules
- AC Inputs for 24 V, 120 V, 240 V
- DC Inputs for 3.3 to 32 V, 10 to 60 V
- UL Recognized (E46203)/CSA Certified (38595)
- 4KV Optical Isolation
- Open-Collector Output
- Industry Standard Packaging

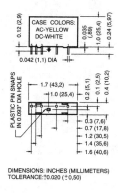

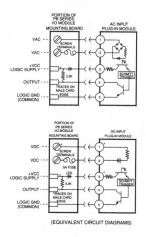

(EQUIVALENT CIRCUIT DIAGRAMS)

Figure 7.9 Digital Input I/O Module *(Courtesy Gordos (Crouzet Corporation))*

INTERFACING PROBLEMS

Problem #1. Wire the necessary circuitry so that an electromechanical relay will operate from one of the output bits from the MetraByte PIO-12 or other digital I/O board. A 120-volt light bulb in an appropriate socket should be connected to the contact side of the relay. **(Use extreme caution as the voltages involved in this activity are dangerous to you and the computer system.)**

Write a program which incorporates the DOS clock so that the light bulb will be ON for 2 minutes and OFF for 1 minute. The program should loop until any key is pressed at which time the program will quit executing. Demonstrate the circuit to your instructor.

SM SERIES OUTPUT MODULES

FEATURES

- UL Recognized (E46203)/CSA Certified (38595)
- AC Modules have High Current Thyristor with 100 Amp Surge Capability
- Zero or Random Turn-On Available in AC Modules
- Plug into Mounting Boards for SM or 0.6" Modules
- 4KV Optical Isolation (1500 VAC Optical Isolation for FET DC Output Modules)
- Industry Standard Packaging
- 3.5 Amp AC Modules Provide Extra Switching Capability
- 5.0 Amp DC Modules Available

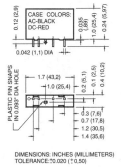

CASE COLORS:
AC-BLACK
DC-RED

PLASTIC PIN SNAPS IN 0.093" DIA HOLE

DIMENSIONS: INCHES (MILLIMETERS)
TOLERANCE:±0.020 (±0,50)

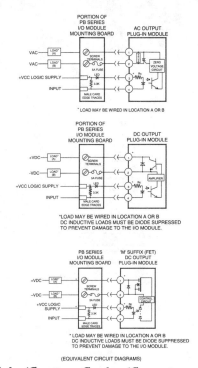

(EQUIVALENT CIRCUIT DIAGRAMS)

Figure 7.10 Digital Output I/O Module *(Courtesy Gordos (Crouzet Corporation))*

Problem #2. Modify the program written for Problem #1 so that the light will be lit for the first 10 seconds of each minute, go OFF for the next 20 seconds, turn ON for the next 10 seconds, turn OFF and remain OFF for the remainder of the minute. The program should loop repeating this pattern until halted by pressing any key on the keyboard. Demonstrate the circuit to your instructor.

Problem #3. Wire the necessary circuitry so that eight DIP switches are connected to an input port and a solid-state relay is connected to one bit of an output port. Wire a 120-volt light bulb as a load to the output side of the SSR. **(Use extreme caution as the voltages involved in this activity are dangerous to you and the computer system.)**

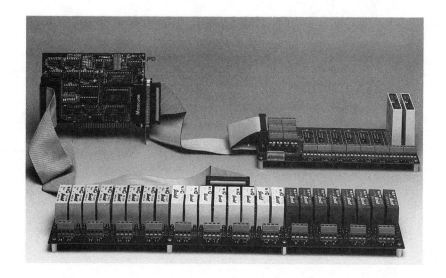

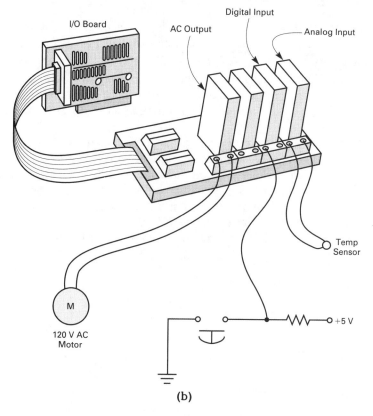

(b)

Figure 7.11 Digital I/O Mounting Rack

Write the necessary program so that the data input from the DIP switches is displayed on the CRT. If the incoming data byte is $\geq$ 0FH, the bulb should be lit. If the incoming byte is $<$ 0FH, the bulb should be OFF. Design the program so that it will loop until a key is pressed on the keyboard. Demonstrate the circuit to your instructor.

Problem #4. Wire the necessary circuit so that a digital input signal of 12 volts can be interfaced into the PIO-12 board. This assignment will require the construction of a circuit similar to that shown in Figure 7.7.

Write the necessary program so that each time the switch is closed the following message appears on the CRT screen.

```
12 volt switch is CLOSED
```

Each time the switch is opened the following message should appear on the CRT screen.

```
12 volt switch is OPEN
```

The program should be written so that only one of the above messages appears on the CRT screen. Demonstrate the circuit to your instructor.

Problem #5. Wire the necessary circuit so that an eight bit DIP switch is connected as an input device to an input port to the computer. Also wire two solid-state relays to an output port so that each relay is controlled by a different bit. Connect a 120 volt light bulb to one of the relays and a 120 volt "muffin" fan to the other relay.

Write the necessary program so that if the incoming data from the DIP switch is greater than 127 decimal the muffin fan will be ON. If the incoming data is less than 127 decimal the light bulb will be ON, and if the data is exactly 127 neither device will be ON. Develop the program so that the fan and light bulb are not both ON at the same time. Also provide a means so that the program may be halted from the keyboard.

This is not a trivial assignment—think in through! Demonstrate the circuit to your instructor.

CHAPTER 8

Interfacing to the PC Bus: Analog Input

ANALOG SIGNALS

Most sensors in industrial and process control applications produce analog electrical signals. This means that the character of the signal is continuous with an infinite number of values as the amplitude of the signal rises and falls. The analog signal produced by a sensor may be a slowly changing DC voltage, possibly varying between 0 and +5 volts, or a varying DC current signal ranging from 4 to 20 mA. The analog signal may also be an AC voltage varying at frequencies ranging from a few hertz to megahertz. Most computers are digital devices, therefore, they are inherently incompatible with the analog signals produced by these "real-world" sensors. An example of a typical signal produced by an analog sensor is the voltage signal produced by the thermistor temperature sensing circuit shown in Figure 8.1.

The thermistor shown in Figure 8.1 is manufactured by Thermometrics, Inc., 808 U.S. Highway 1, Edison, NJ 08817. This thermistor is rated at 10 K ohms at 25 degrees Centigrade. As the thermistor changes resistance in response to variations intemperature, the circuit produces a variable DC voltage, the value of which is dependent upon the ambient temperature. Most thermistors have a negative temperature coefficient, thus its resistance will decrease as the temperature surrounding the thermistor increases. Conversely, the resistance of the device will increase as the temperature falls. Since the power source for the thermistor circuit is a DC voltage, the signal output from the sensing circuit is a varying DC voltage that rises or falls as the temperature varies. This varying analog voltage represents the temperature surrounding the sensor. Observe that the output voltage from the sensor circuit (V_O) increases as the temperature increases because of the relationship between the sensor and the series resistor.

ANALOG-TO-DIGITAL CONVERSION

To allow the analog signal from the circuit of Figure 8.1 to be processed by a computer, the analog voltage must be converted to a digital signal that is compatible with the TTL voltage ranges required internally by personal computers. The conversion of the analog signal to a digital format is accomplished by an *analog-to-digital* (A/D) *converter*. Conversion of the analog signal into an equivalent digital value is accomplished by an integrated circuit chip on the A/D data acquisition and control board. A block diagram of a representative A/D converter is shown in Figure 8.2. The A/D converter digitizes the incoming analog signal providing an output which is a digital representation of the original analog voltage.

The A-to-D converter can be thought of as a device that "slices" the span or range of the incoming analog voltage into slices or steps. The thickness of the "slices" that the incoming analog signal is divided into is referred to as the *resolution* of the particular chip and A/D board. Resolution depends on the number of binary bits (n) that the converter uses in the conversion process. Contemporary A/D converters use eight bits, 10 bits, 12 bits, 14 bits, or 16 bits to represent the analog input voltage. The converter "slices" the analog value into $2^n - 1$ steps or parts. As the number of binary bits increases, the "thinner" each slice becomes. An increase in resolution means that more subtle variations in the analog signal can be detected by the binary computer. The cost of an analog-to-digital DA & C board is primarily a function of the number of bits in the A/D converter. Many popular A/D converters use eight or 10 bits as a compromise between the higher cost and correspondingly more accurate readings of 12-bit, 14-bit, and 16-bit converters. As with many choices in life, economics plays a role in the selection of A/D data acquisition products.

Resolution of the 8-bit A/D converter shown in Figure 8.2 can be calculated based on the maximum range of input analog voltage (span) that will be applied to the converter and the number of binary bits that provide the resultant binary output value. The converter represented in Figure 8.2 is typical of many A/D converters in that the converter is capable of accepting input voltages that fall within the range of -5 to $+5$ volts. If the bipolar span of -5 to $+5$ volts is not suitable, other converters are available that will accept different voltage ranges. Common input voltages for A/D converters are 0 to $+5$ volts, 0 to $+10$ volts, and -10 to $+10$ volts. Referring to the converter shown in Figure 8.2, the following formula will show how the resolution for that converter is calculated. This calculation considers the voltage span and the 8-bit digital output from the converter.

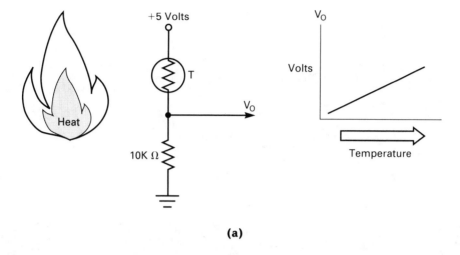

(a)

Thermometrics, Inc. D320, D200, and D120 series thermistors are very low cost, large size, epoxy coated disc devices with greater power handling capabilities than series C100C Chips making them ideally suited for many temperature compensation, current limiting, and delay circuit applications.

These NTC devices are available in a wide range of resistance values @ 25° C ranging from 13 ohms to 400 kilohms with tolerances of either ± 15% or ± 20% depending on devices selected.

For more information or for assistance with your application, please contact Thermometrics Applications Engineering Department.

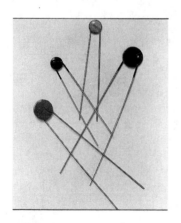

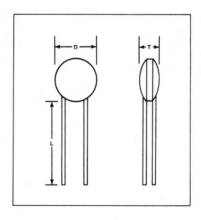

(b)

Figure 8.1 Thermistor Sensor and Application Circuit *(Courtesy Thermometrics, Inc.)*

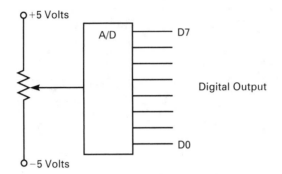

Voltage	Binary Output	
−5	0000 0000 ◄	
0	1000 0000	10 Volt Span
+5	1111 1111 ◄	

Figure 8.2 Block Diagram of A/D Converter

$$\text{Step Size} = (V_{max} - V_{min})/(2^n - 1)$$

$$\text{Step Size} = (+5 - (-5))/255$$

$$\text{Step Size (resolution)} = .0392 \text{ volts}$$

The calculation shows that for a span of 10 volts and eight bits of digital output the resolution is approximately 39 mV.

It is important to know the resolution of any A/D board used in an interfacing project. The implication of knowing the resolution, or step size, is that analog voltage changes that are smaller than the resolution value will not be detected by the A/D converter. Therefore, a change in input signal will not be communicated digitally to the computer. If increased sensitivity is necessary, a converter with greater resolution can be used. Applying the previous calculation to an A/D converter that uses 12 binary bits shows that a significant improvement in resolution sensitivity can be obtained by using a converter with more digital bits. The following formula calculates the resolution for a 12-bit A/D converter with a −5 to +5 volt signal applied.

$$\text{Step Size} = 10 \text{ v}/(4096 - 1) = 2.44 \text{ mV}$$

Users of A/D systems must weigh the importance of the resolution as compared to the cost of the converter. This decision will influence which A/D converter will be used in an interfacing application.

TYPES OF ANALOG-TO-DIGITAL CONVERTERS

Three types or configurations of analog-to-digital converters are commonly used on DA & C boards. The differences among the three converters is the technique used by the chip to convert the analog value to its digital equivalent as well as the amount of time required for the conversion.

The simplest and least expensive A/D converter is the *dual-slope converter*. This converter is relatively slow, requiring approximately 10 to 100 ms for the conversion process to be completed. An advantage of this converter is its low sensitivity to electrical noise and its immunity to minor variations in the analog input signal. The Dual-Slope converter is also the least expensive analog-to-digital converter. In spite of its low cost and high noise immunity, the Dual-Slope converter is not used extensively in real-time process control applications because of its slow conversion rate.

The most popular type of analog-to-digital converter is the successive-approximation converter. This converter is popular due to its faster conversion speed and modest cost. An informal survey of A/D converter boards indicates that most converter boards use the successive-approximation method of conversion. Figure 8.3 shows an 8-bit analog-to-digital converter board produced by Keithley MetraByte Corporation.

This A/D converter board, identified as the Model DAS-4, is a low cost, yet highly versatile, A/D board that is suitable for most analog applications. This board is rated for analog signals ranging from -5 volts to $+5$ volts. This input voltage span in conjunction with the 8-bit converter chip on the board yields a resolution of 39 mV.

The *Flash A/D Converter* has been designed for applications that require extremely fast A/D conversions. This converter uses a large number of analog comparators in its internal circuitry. Comparators do not require clock signals to "step through" the conversion process; therefore, the conversion process is accomplished much quicker than with the successive-approximation form of converter. The conversion of an analog signal to its digital equivalent in a flash converter is a continuous ongoing process requiring approximately 10 ns to complete a conversion. Flash converters are expensive, and the level of sophistication of the application should justify the cost of using these "high-end" devices.

SAMPLE-AND-HOLD CIRCUITS

Most A/D interface boards incorporate a *sample-and-hold* (S/H) *circuit* on the DA & C board. The function of this circuit is to stabilize the changing analog signal while the conversion process is taking place. A typical technique used in these boards to "freeze" the analog voltage is to include a capacitor on the board which

FEATURES

- 8 A/D channels with 8-bit resolution
- Extremely low cost
- 7 digital I/O bits (4 output, 3 input)
- ±5 volt inputs (for 39 mV resolution)
- Interrupt handling capability
- Software and complete Users Manual included

APPLICATIONS

- Data logging
- Alarm condition monitoring
- Laboratory automation
- Student experiments
- Home experiments
- Energy management

DAS-4's fixed ±5 volt full-scale input range provide a minimum resolution of .039 volts (39 millivolts). The 8 inputs are single-ended (share a common ground or return line) and are protected from overvoltages up to ±30 volts.

A complete utility software package is included with the DAS-4 and contains a Mode Call Driver that greatly simplifies programming. The package also includes example programs, initialization and set-up routines and a simple strip-chart recorder emulation routine.

Seven bits of TTL/CMOS-compatible digital I/O are provided on the board, 4 digital outputs and 3 digital inputs. Each output will handle 5 standard TTL loads and can sink 8 mA. An external interrupt input is provided that is jumper-selectable to any of the PC interrupt levels (2 – 7). This allows the DAS-4 to sample data triggered by an external digital signal. PC-bus power (+5, and ±12V) is also provided on the DAS-4's connector, allowing external circuitry to be operated from the computer's power supplies.

FUNCTIONAL DESCRIPTION

Keithley MetraByte's DAS-4 is an extremely low cost, yet powerful analog input board for the IBM PC/XT/AT and compatibles. The board plugs directly into one of the I/O expansion slots within the computer. A 37-pin "D" connector extends outside the computer and provides the connection to external wiring.

The DAS-4 includes 8 analog input channels with an 8-bit successive approximation A/D converter. The 8-bit resolution combined with the

BLOCK DIAGRAM

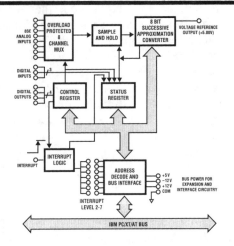

Figure 8.3 Successive-Approximation A/D Converter Board *(Courtesy MetraByte Corporation)*

is charged to the value of the incoming analog voltage. The capacitor provides a stable analog voltage to the converter while the conversion process is underway. Figure 8.4 shows in block diagram form the internal architecture of the MetraByte DAS-4 A/D board. Note that the sample-and-hold portion of the circuit is placed before the analog-to-digital converter module. This location is consistent with the role of the S/H circuit.

Another feature typical to many A/D converter boards is the availability of multiplexed input channels. This feature, found on the DAS-4 board, is shown in the block diagram of Figure 8.4. The DAS-4 has eight single-ended input channels which are multiplexed to a single 8-bit successive-approximation converter. The ability to use more than one analog input is an attractive feature to users of this board. For the price of the DAS-4, the user obtains not one A/D converter but effectively eight converters. Each channel must be accessed individually. Only one channel can be read by the converter and converted at a time.

THE DAS-4 ANALOG-TO-DIGITAL CONVERTER

The MetraByte DAS-4 is a low cost, 8-bit, 8-channel, analog-to-digital converter that is capable of accommodating analog inputs within the range of −5 to +5 volts. This board is installed into the computer like other DA & C boards. As with other I/O boards, the base address for the A/D board is set by DIP switches that are located on the board. Remember that *all power to the computer must be OFF when installing or removing any computer board!*

Figure 8.5 illustrates the base address DIP switch for the DAS-4. The selection of the base address must be chosen so that it does not conflict or overlap with the addresses used by other boards. Refer to the I/O port map shown in Figure 5.3 for further information on available addresses allocated to all types of I/O boards. For purposes of this illustration, a base address of 310 Hex was chosen for the DAS-4. This address will be used in software examples and when providing sample programs for this board. The address was selected so that it would not conflict with the range of addresses required by the digital I/O board which was previously assigned a base address of 300 Hex. The designated base address of 310 Hex is not the only address available. Any other, nonconflicting, address could be selected. Based on the range of available addresses identified in Figure 5.3 it is recommended that addresses within the range of 300 Hex to 31F Hex be selected.

PROGRAMMING THE DAS-4

Two major programming techniques are available to the user of the DAS-4. Users may elect to control the A/D board directly through writing their own software routines or through subroutine software supplied by Keithley MetraByte, or third-

BLOCK DIAGRAM

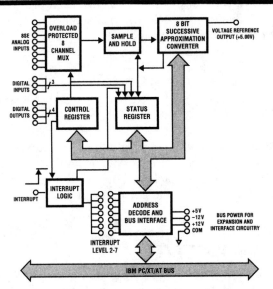

Figure 8.4 DAS-4 Block Diagram *(Courtesy Keithley MetraByte Corporation)*

party vendors. A disk containing BASIC-callable subroutines written by Keithley MetraByte Corporation comes with each DAS-4 board at the time of purchase. This disk is labeled **DAS-4 Utility Software**. Use of the DAS-4 is not dependent upon the subroutines on this disk. The subroutines are provided as a software development time-saving option that the user may elect to use.

This text will describe both programming options; the development of user written software and the use of supplied subroutines. The reader must keep one important condition in mind; the subroutines supplied by MetraByte for the DAS-4 are not compatible with the C language. The subroutines provided on the disk are accessible only through BASIC! For those readers familiar with assembly language programming there is also an assembly source file on the disk supplied by Keithley Metrabyte. If the user is an experienced assembly/C programmer this file can be modified to be used with C.

If the user wants to control the DAS-4 directly by writing all of the software, the initialization process follows a procedure similar to that used with the PIO-12 digital I/O board. A specific bit pattern must be written to a *control register*. The 8-bit control register is addressed at base address + 2. For purposes of the example used in this chapter, the control register has an address of 312 Hex. Writing a byte to the control register selects the multiplexed channel to be read, enables or disables the interrupts, and provides the digital values that the DAS-4 can output through

Specifying the Base Address

The Base Address switch is preset at the factory for 300-Hex.
   If this address is already assigned to some other device in
your computer, you must set the Base Address switch to specify
a different address.

BASE ADDRESS

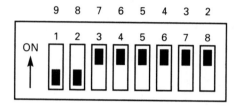

Switch	Address line	Value when Switch is OFF	
		Decimal	Hex
1	A9	512	200
2	A8	256	100
3	A7	128	80
4	A6	64	40
5	A5	32	20
6	A4	16	10
7	A3	8	8
8	A2	4	4

Figure 8.5 DAS-4 Base Address DIP Switch

its binary I/O bits. Figure 8.6 provides information about the role of each of the
eight bits in the control word. As is typical with most control registers, this is a
write only register.

   The following definitions or descriptions apply to the bits in the control register.

OP4-OP1. These bits correspond to the four general purpose digital outputs
available. These lines can be used for external control.

DAS-4 Control Register

D7	D6	D5	D4	D3	D2	D1	D0
OP4	OP3	OP2	OP1	INTE	MA2	MA1	MA0

(Base Address +2)

Channel Select Code

MA2	MA1	MA0	Channel
0	0	0	0
0	0	1	1
0	1	0	2
0	1	1	3
1	0	0	4
1	0	1	5
1	1	0	6
1	1	1	7

Figure 8.6 DAS-4 Control Register

INTE. Interrupts are disabled if this bit is made low.

MA2-MA0. These three bits are used to select the analog channel to be connected to the sample and hold section of the DAS-4. In other words, the binary value in these bits selects which analog channel to be read.

During power-up the RESET line of the computer is asserted. This causes the control register of the DAS-4 to be cleared. Clearing the register insures that the DAS-4 interrupts are disabled, all digital outputs are low, and that analog channel 0 is selected as the default input channel. If the programmer wants to accomplish the same resetting function through software, the following statements in C or BASIC can be executed. Either statement writes eight 0s in binary to the control register.

C Version

```
out(0x310,0);        /* In C */
```

BASIC Version

```
OUT &H310,0          ' In BASIC
```

By way of comparison, to select channel 7 as the active channel for incoming analog voltage, to set all of the digital outputs to a logic 1, and to enable interrupts, the data value sent to the control register would be FF Hex or decimal 255.

Following initialization of the control register the DAS-4 must be instructed to begin the conversion process. This is accomplished by writing **ANY** numerical value to the board's base address. The value written is not important, it is the act of writing to this address that starts the conversion process. Completion of the conversion process is "announced" to the computer by a logic change on one bit of the board's status register. One way of detecting this change is through software. The status register is a **read only** register with an address referenced to the base address of the board. The address of the status register is base address + 3. Figure 8.7 provides details about the significance of each bit in the status register.

> **EOC.** End-of-conversion. If this bit is a high (logic 1), the A/D converter is busy performing a conversion. This bit will be reset to a logic 0 at the completion of the conversion process.
>
> **IP3–IP1.** These three bits correspond to the three digital inputs available on the DAS-4. These lines can be used to monitor digital devices.
>
> **IRQ.** If interrupts have been enabled, this bit will be set to a logic high when an interrupt has been generated. A write operation to the control register resets the interrupt and forces this bit to a logic low.
>
> **MA2–MA0.** These three bits simply identify which channel is currently selected. In other words, when finding out if EOC is low, the program also can tell which channel has just been converted.

It is important to remember that the status register is a read only register and that all changes in channel selection must be made through the control register.

The EOC bit can be monitored simply by reading the status register and masking-out all bits except the EOC bit. If the EOC bit is not logic 0, then the program should loop until it is logic 0. The following lines of code are examples

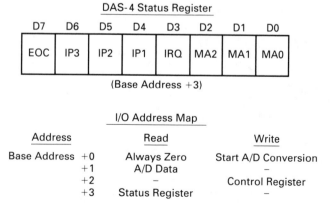

DAS-4 Status Register

D7	D6	D5	D4	D3	D2	D1	D0
EOC	IP3	IP2	IP1	IRQ	MA2	MA1	MA0

(Base Address +3)

I/O Address Map

Address	Read	Write
Base Address +0	Always Zero	Start A/D Conversion
+1	A/D Data	–
+2	–	Control Register
+3	Status Register	–

Figure 8.7 DAS-4 Status Register Bits

of a group of instructions that can be used to check the end-of-conversion bit. Each of the programs will loop until the conversion is completed:

C Version

```
do                         /* Top of Loop           */
   stat = inp(0x313);      /* read the Status Reg    */
while((stat & 0x80) = 0x80);  /* Wait till bit is 0 */
```

BASIC Version

```
DO                              ' TOP OF LOOP
   STAT-INP(&H313)              ' READ THE STATUS REGISTER
LOOP WHILE((STAT AND &H80) = &H80)' WAIT TILL EOC BIT IS 0
```

Once the conversion is completed and the EOC bit is toggled low, the converted data, in binary form, is available to be read by the computer. The binary data is located in a register with the address of base + 1. In all the examples for the DAS-4 that we've discussed so far we have used a base address of 310 Hex; therefore, the data will be read from the register located at address 311 Hex.

Figure 8.8 illustrates some sample readings that are the result of applying an increasing analog voltage to the DAS-4 analog-to-digital converter. A check of this figure will confirm that a -5 volt analog input signal equates to 00 Hex and that spanning the nearly 10 volt range to $+4.96$ (roughly $+5$) volts yields a digital value of 255 decimal or FF Hex.

Binary	Hex	Decimal	Analog Input Voltage
0000 0000	00	0	-5.000 v ($-$Full scale)
0000 0001	01	1	-4.961 v
. .	.	.	.
0100 0000	40	64	-2.500 v ($-1/2$ scale)
. .	.	.	.
1000 0000	80	128	$+/-0$ v (zero)
1000 0001	81	129	$+0.039$ v
. .	.	.	.
1100 0000	C0	192	$+2.500$ v ($+1/2$ scale)
. .	.	.	.
1111 1111	FF	255	$+4.961$ v ($+$Full scale)

Figure 8.8 Sample Binary Output Values from DAS-4

The program shown in Figure 8.9 contains all the instructions needed to perform a single analog-to-digital conversion on the analog voltage present in the schematic shown in Figure 8.1 and to print the result on the computer's CRT screen. The result will be displayed as both a decimal value and as a voltage value on the CRT.

Conversion of the decimal value to a voltage value is done by a formula in the software. The formula *(((d − 128) ∗ 5) / 128)* is used to convert the digital value from the DAS-4 to a floating point number that corresponds to the voltage at the input. This formula takes into account that the circuit's supply voltage is +5 volts and that the span of the incoming analog voltage is limited to the range of 0 to +5 volts. As the thermistor cools the voltage applied to the converter, the value displayed on the CRT will fall to value closer to 0 volts. As the temperature increases, the voltage value applied to the converter and displayed on the CRT will rise closer to the upper limit of +5 volts.

For example, if the DAS-4 returned a digital 230 the formula would be *((230 − 128) ∗ 5) / 128) = 3.98 volts*. A full scale output reading from the board

```
#include <stdio.h>
#include <dos.h>                     /* Needed for inp() and outp()       */
#include <conio.h>                   /* Needed for gotoxy()               */
#define BASE 0x310                   /* Board Base address                */
#define DATA BASE+1                  /* Data Reg address                  */
#define CTRL BASE+2                  /* Control Reg address               */
#define STATUS BASE+3                /* Status Reg address                */

void main(void)
  {
    unsigned char stat, d;
    float Volts;
    clrscr();                        /* Clear the screen                  */
    outp(CTRL,0);                    /* Write Control Word                */
    outp(BASE,0);                    /* Start the conversion              */
    do                               /* Loop ...                          */
      stat = inp(STATUS);            /*   by getting the status           */
    while((stat&0x80)==0x80);        /* Until End Of Conversion           */
    d = inp(DATA);                   /* Read the data                     */
    Volts = (((float)d - 128) * 5 ) / 128);     /* Calculate volts
    gotoxy(20,7);                    /* Set screen position               */
    printf("The raw value was %u",d);   /* Print it                      */
    printf(" and that calculates to be %.2f Volts\n",Volts);
  }




'
'
'
'
'
CONST BOARD = &H310
CONST DAT = BOARD + 1
CONST CTRL = BOARD + 2
CONST STATUS = BOARD + 3
CLS                                        ' CLEAR THE SCREEN
OUT CTRL, 0                                ' WRITE CONTROL WORD
OUT BOARD, 0                               ' START THE CONVERSION
SLEEP (2)                                  ' SHORT DELAY
DO                                         ' LOOP ...
  STAT = INP(STATUS)                       '    GETTING STATUS
LOOP WHILE ((STAT AND &H80) = &H80)        ' UNTIL END OF CONVERSION
D% = INP(DAT)                              ' READ THE DATA
VOLTS = (D% - 128) * 5 / 128               ' CALCULATE VOLTS
LOCATE 7, 20                               ' SET SCREEN POSITION
PRINT "The raw value was "; D%;
PRINT " and that calculates to be "; VOLTS; " Volts"
END
```

Figure 8.9 A/D Conversion Program

of a decimal 255 would yield the following results; *(((255 − 128) * 5) / 128) = 4.96 volts*. Either of these examples might be a typical voltage reading from the thermistor circuit in Figure 8.1.

Using a DAS-4 Driver

If you are programming in BASIC and plan to use the driver supplied by Keithley MetraByte, you should follow the appropriate directions supplied with the software to install and link the driver to the programs and examples that follow.

The DAS-4 driver is a group of subroutines that perform specific functions including initialization, conversion control, and other functions. Figure 8.10 contains a listing of the 10 "modes" or subroutines available with this software. These subroutines are accessed through BASIC's CALL command.

To use the driver, the first step is to initialize the DAS-4 driver (mode = 0). A call to this subroutine tells the driver the address of the board and the level of interrupts to be generated. Once the board has been initialized, the channel or channels that will be used are selected. This selection process is shown in Figure 8.11. In this example, the channel selected for use is channel 4. This is done by setting both the high and low scan limits to 4 through mode 1. Finally, a call to mode 2 will do one A/D conversion and return the decimal value to the CRT screen.

If the BASIC driver is used and the programmer desires to calculate the analog voltage applied to the DAS-4, a special formula must be used. The program previously shown is not compatible with the driver. If the driver supplied with the DAS-4 is used, the decimal value returned after the conversion is in the range of −128 to +127. The following example illustrates some digital values that are returned from the converter and the corresponding analog input voltages.

Digital Value	Analog Input
−128	−5.00 v
−127	−4.96 v
...	...
...	...
0	0.00 v
+1	+0.039 v
...	...
...	...
127	+4.96 v

Figure 8.11b is a straight forward example of how to use the DAS-4 driver from BASIC. If the reader is using Quick BASIC, refer to the section in the DAS-4 manual that describes creating a Quick Library. After the library has been created,

Examples of the Use of the Call Routine

The following subsections give detail information and examples of the use of the CALL routine in all 10 modes. The modes are selected by the MD% parameter in the CALL as follows:-

Mode		Function
0	. . .	Initialize, input DAS-4 base address, interrupt level & check hardware
1	. . .	Set multiplexer low & high scan limits
2	. . .	Perform a single A/D conversion. Return data and increment multiplexer address. (Programmed conversion). Speed up to 200Hz, Operation—foreground.
3	. . .	Perform an N conversion scan after trigger. Conversions initiated by an external input. Data is transferred to an integer array. Speed up to 3KHz. Operation—foreground.
4	. . .	Perform an N conversion scan after trigger into a memory buffer area. Conversions initiated by an external input. Data transferred by interrupt. Speed up to 3KHz. Operation—background.
5	. . .	Analog trigger function similar to a scope trigger (specify channel, level & slope)
6	. . .	Transfers data from memory buffer to an integer array either as a whole block or piece by piece (used after mode 4).
7	. . .	Return status. Reports next channel number to be converted, whether interrupt is active or finished, interrupt level and remaining number of conversions in interrupt mode 4.
8	. . .	Read digital inputs IP1-3.
9	. . .	Write digital outputs OP1-4.

Figure 8.10 DAS-4 Modes *(Courtesy Keithley MetraByte Corporation)*

you must start the Quick BASIC program with the command **QB /L DAS4.QLB**. This command causes Quick BASIC to execute and to include the additional library file **DAS4.QLB** when the program is created.

Wiring Connections

Connecting analog inputs or digital I/O to the DAS-4 is similar to the techniques used with the digital PIO-12 board. All connections to the DAS-4 are via a 37-pin D-type male connector which is part of the A/D board. A 37-conductor ribbon cable can be used to connect the DAS-4 to the universal screw terminator board,

```c
#include <stdio.h>
#include <conio.h>                     /* Needed for gotoxy()          */
#define BASE 0x310                      /* Set Board base address       */

extern void das4(int,int [],int *);    /* Prototype for driver         */

int md, flag,d[4];                      /* These MUST be global         */

void main(void)
  {
    float Volts;

    clrscr();
    md=0;                               /* Mode 0 - INITIALIZATION      */
    d[0]=BASE;                          /* Board Address                */
    d[1]=7;                             /* Interrupt Level  2 - 7       */
    das4(md,d,&flag);                   /* Call driver                  */
    if(flag!=0)                         /* Check return code   0 = OK   */
       {
         printf("Error Initializing\n");
         exit(flag);
       }

    md=1;                               /* Mode 1 - Set scan limits     */
    d[0]=4;                             /* Lower limit                  */
    d[1]=4;                             /* Upper Limit                  */
    das4(md,d,&flag);                   /* Call driver                  */
    if(flag!=0)                         /* Check return code   0 = OK   */
       {
         printf("Error setting limits\n");
         exit(flag);
       }

    md=2;                               /* Mode 2 - Do one converssion  */
    das4(md,d,&flag);                   /* Call driver                  */
    if(flag==0)                         /* Check return code   0 = OK   */
       {
         gotoxy(20,7);                  /* Set screen position          */
         printf("The raw value was %d",d[0]);
         Volts = (float)d[0] * 5 / 128;
         printf( " and that calculates to be %.2f Volts\n",Volts);
       }
  }
```

(a)

```
'
'
'
'
'
CONST BOARD = &H310
CLS                                     ' CLEAR THE SCREEN
MD% = 0                                 ' MODE 0 - INITIALIZATION
D%(0) = BOARD                           '    BOARD ADDRESS
D%(1) = 7                               '    INTERRUPT LEVEL  2 - 7
DAS4(MD%, D%(0), FLAG%)                 ' CALL DRIVER
IF( FLAG% <> 0)                         ' CHECK RETURN CODE  0 = OK
  PRINT "Error Initializing"
  STOP
END IF

MD% = 1                                 ' MODE 1 - SET SCAN LIMITS
D%(0) = 4                               '    LOWER LIMIT
D%(1) = 4                               '    UPPER LIMIT
DAS4(MD%, D%(0), FLAG%)                 ' CALL DRIVER
IF( FLAG% <> 0)                         ' CHECK RETURN CODE  0 = OK
  PRINT "Error Setting limits"
  STOP
END IF

MD% = 2                                 ' MODE 2 - DO ONE SCAN
DAS4(MD%, D%(0), FLAG%)                 ' CALL DRIVER
IF( FLAG% = 0)                          ' CHECK RETURN CODE   0 = OK
  LOCATE 7,20                           ' SET SCREEN POSITION
  PRINT "The raw value was ";D%(0);
  VOLTS = D%(0) * 5 / 128
  PRINT " and that calculates to be ";VOLTS;" Volts"
END IF
END
```

(b)

Figure 8.11 Sample Driver Call

Model STA-U, described earlier or to another form of terminal block. The STA-U accessory is an ideal terminal block that allows the orderly connection of wires to the DAS-4. Figure 8.12 shows the connector pin assignments on the DAS-4 and the manufacturer's specifications for the board.

DAS-8 ANALOG-TO-DIGITAL CONVERTER

In many applications, the resolution associated with the DAS-4 analog-to-digital converter is not accurate enough. These situations require a more precise analog reading than the 39 mV offered by the 8-bit DAS-4. The DAS-8 A/D converter, manufactured by Keithley MetraByte Corporation, is suggested as a way of meeting the need for higher resolution. The DAS-8 is a 12-bit successive-approximation converter with a resolution of 2.44 mV. This product, like other A/D boards, has eight multiplexed channels.

The specifications indicate that the DAS-8 has an input range of -5 volts to $+5$ volts and an input impedance of 10 Meg ohms. The input voltage range of this product makes it compatible with sensors and signal conditioning circuits that were used with the DAS-4. The same electrical signals that were sent to the DAS-4 may be connected to the DAS-8. The DAS-8 will provide significantly improved measurements.

Just like other DA & C boards, the DAS-8 is controlled through internal registers. Figure 8.14 provides a list of the registers and their function.

CONNECTOR PIN ASSIGNMENTS

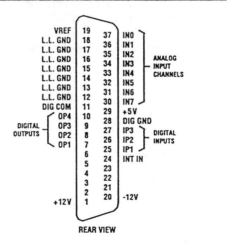

Figure 8.12 DAS-4 Pin Assignments *(Courtesy Keithley MetraByte Corporation)*

DAS-8 SPECIFICATIONS

Power Consumption

+5 V supply	107 mA typ./180 mA max.
+12 V supply	6 mA typ./10mA max.
−12 V supply	10 mA typ./16 mA max.

Analog Inputs

Number of Channels	8, Single-ended
Resolution	12 bits. (2.4 mV/bit)
Accuracy	0.01% of reading ± bit.
Full scale	±5 Volt
Coding	Offset binary
Overvoltage	Continuous single channel to ±35 V
Input current	100 nA max at 25° C.
Gain Tempco	Gain or F.S., ±25 ppm/°C max.
	Zero, ±10 microvolt/°C max.

A/D Specification

Type	Successive approximation.
Resolution	12 bit
Conversion	25 us typ. (35 us max.)
Monotonicity	Guaranteed over operating temperature range.
Linearity	±1 bit.
Zero drift	10 ppm/°C max.
Gain drift	50 ppm/°C max. (30 ppm/deg C.,opt.)

Sample Hold Amplifier

Acquisition time	15 ms to 0.01%, typical for full-scale step input
Dynamic	1 bit (2.44 mV) @ 2000 V/sec sampling error

Figure 8.13 DAS-8 Specifications *(Courtesy Keithley MetraByte Corporation)*

Most of the registers and their functions are similar to the registers found in the DAS-4. Any slight differences between the two converters will be illustrated in Figure 8.15.

Notice in Figure 8.15 that there is a shifting of data within the program. The shifting is necessary because of the way in which the 12-bits are provided by the converter and the bit size of a memory location. Since a memory location can hold one byte of data, it is clear that more than one memory location is needed to store the 12 bits that are provided by the converter. Assembling of the 12 bits is illustrated in Figure 8.16.

This figure shows that the *least significant byte* (LSB) of the data is stored in a register having the base address of the A/D board. The *most significant byte*

ADDRESS		READ	WRITE
Base Address	+ 0	A/D Lo Byte	Start 8 bit A/D Conv
	+ 1	A/D Hi Byte	Start 12 bit A/D Conv
	+ 2	STATUS reg	CONTROL reg
	+ 4	Read Ctr 0	Load Ctr 0
	+ 5	Read Ctr 1	Load Ctr 1
	+ 6	Read Ctr 2	Load Ctr 2
	+ 7	-	Counter Control

Figure 8.14 DAS-8 Registers

(MSB) is stored at an address equal to base address + 1. These two bytes are shifted and ADDed producing a word consisting of the 12 bits.

The LSB is shifted to the right four places while the MSB is shifted to the left four places. These two values are ADDed producing the 12-bit output. This technique is typical of the bit manipulation that is necessary when A/D converters have a resolution greater than one byte.

An attractive feature of the DAS-8 is that Keithley MetraByte Corporation produces an optional driver for this product that allows calls from the C language. The company offers this software product because of the immense popularity of the DAS-8. This board, which combines low cost and the accuracy of 12-bit resolution, is very popular for applications in which resolution is important. An application in which this accuracy is critical is when a thermocouple is used to measure temperature. The nature of thermocouples are that they produce a very small output voltage, generally in the magnitude of microvolts. As the temperature changes the corresponding change in the thermocouple output voltage is very small. The resolution and accuracy offered by the 12-bit A/D converter is necessary to use the thermocouple effectively.

Contec Microelectronics U.S.A. Inc., 2188 Bering Drive, San Jose, CA 95131 offers an analog-to-digital converter board that is similar to the DAS-8. The board produced by Contec is shown in Figure 8.17. This board offers eight multiplexed input channels that can be configured for −5 to +5 volts bipolar or 0 to +10 volts unipolar.

The manufacturer of this board, like manufacturers of competing products, offers a BASIC software driver as a standard item with the purchase of the board. An optional C driver is available as are drivers for PASCAL and Microsoft Assembler.

Most I/O board manufacturers produce at least one specialized thermocouple input board. The board shown in Figure 8.18 is representative of this type of board.

This board, manufactured by Industrial Computer Source, 10180 Scripps Ranch Blvd., San Diego, CA 92131, is identified as the model PC-73. Designed to serve as an interface between industrial thermocouples and a PC, the board has eight input channels that are compatible with all types of thermocouples. This compatibility includes thermocouple types J, K, and T which are described in

```c
#include <stdio.h>
#include <dos.h>                         /* For inp() and outp()   */
#include <conio.h>                       /* For gotoxy() & clrscr()*/

#define BASE   0x310                     /* Board Base address     */
#define DATA   BASE + 1                  /* Data Reg Address       */
#define CTRL   BASE + 2                  /* Control Reg Address    */
#define STATUS BASE + 3                  /* Status Reg Address     */

void main(void)
  {
    char stat, d;
    float Volts;

    clrscr();
    outp(CTRL,0x00);                     /* Write Control Word     */
    outp(BASE,0x00);                     /* Start the conversion   */
    do                                   /* Loop                   */
      stat = inp(STATUS);                /* by getting the status  */
    while((stat & 0x80) == 0x80);        /* Until End of Conversion*/
    d = inp(DATA);                       /* Read the data          */
    Volts = (((d - 128) * 5 ) / 128); /* Calculate volts        */
    gotoxy(20,7);                        /* Set screen position    */
    printf("The raw value was %d",d);/* Print it               */
    printf(" and that calculates to be %f Volts\n",Volts);
  }
```

```
'
'
'     Example 8.15
'
'
CONST BOARD = &H310                    ' BOARD BASE ADDRESS
CONST DAT = BOARD + 1                  ' DATA REG ADDRESS
CONST CTRL = BOARD + 2                 ' CONTROL REG ADDRESS
CONST STATUS = BOARD + 3               ' STATUS REG ADDRESS

UOT CTRL,&H0                           ' WRITE CONTROL WORD
OUT BOARD,&H0                          ' START THE CONVERSION
DO                                     ' LOOP
  STAT = INP(STATUS)                   '  BY GETTING THE STATUS
LOOP WHILE ((STAT AND &H80) = &H80)' UNTIL END OF CONVERSION
D = INP(DAT)                           ' GET THE DATA
VOLTS = (((D - 128) * 5 ) / 128 )
LOCATE 7,20                            ' SET THE SCREEN POSITION
PRINT "The raw value was ";D;          ' PRINT IT
PRINT " and that calculates to be ";VOLTS;" Volts"
END
```

Figure 8.15 DAS-8 Programming Example

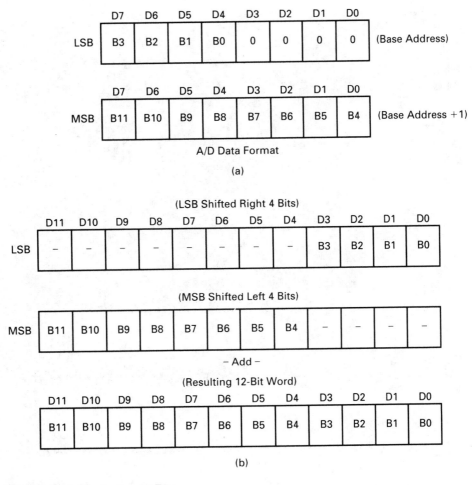

Figure 8.16 Producing 12-Bits

Chapter 3. To simplify the interfacing with a thermocouple, the board is equipped with a provision for a cold junction compensation accessory and a screw connection terminal block. The board is supported by driver software written in both C and BASIC. This software includes driver calls and linearization algorithms. The resolution of this analog input board is 12 bits which is compatible with the accuracy typically needed with thermocouples.

INTERFACING PROBLEMS

Problem #1: Wire the circuit shown in Figure 8.1. Use a +5 volt power supply and a thermistor having similar characteristics to the one shown in the diagram.

Figure 8.17 ADC 30 Analog-to-Digital Converter (Courtesy Contec Microelectronics U.S.A., Inc.)

Write a program that will display the decimal value of the analog signal on the computer screen. Cautiously heat the thermistor with a light bulb, match, or other heat source and observe the rise in the digital value displayed on the screen. Cool the thermistor and note the decreasing change in digital value. Demonstrate the operation of this circuit to your instructor.

Problem #2: Modify the software for Problem #1 to include a formula that will determine the actual analog voltage being applied to the A-to-D board in the

Precision Thermocouple Input Board

Model PC-73

FEATURES
- **8 Differential T/C Input Channels**
- **Use With K, J, E, T, B, R, S & N Thermocouples**
- **Cold Junction Compensation and Screw Terminals on Auxiliary Board**
- **30Hz Data Acquisition**
- **Selectable Gain Control**
- **Supplied With Software**

Description
The PC-73 is a complete interface system consisting of two boards: a half size, PC plug-in board and an auxiliary screw terminal/cold junction compensation board which connects to the PC plug-in board via a ribbon cable which is supplied with the boards.

Driver Software
The driver software allows programmers to control the PC-73 via high level function calls, enabling users to write custom software without needing to understand the low level operation of the card. Advanced multi-segmented polynomial linearization for J, K, E, T, B, R, S and N

type thermocouples is provided. The complete source code, in C and BASIC, is also included for the entire driver package., allowing users to modify existing code, rather than having to start from scratch.

Specifications

Number of Input Channels
8 thermocouple inputs, ungrounded
1 CJC input from terminal board

Resolution
12-bit + sign, 1 in 8192

Total System Accuracy
±¹/₂LSB

Non-linearity
±¹/₂LSB

Effective System Accuracy (Type K)
±0.2°C > -100°C
±0.5°C < -100°C
±2.0°C < -200°C

Input Impedance
1 MOhm/50pF chan off
1 MOhm/200pF chan on

Input Ranges
-4.095 to +4.095 V,
-40.95 to +40.95 mV,
-20.475 to +20.475 mV,
-8.190 to +8.19 mV
Jumper Selectable

Data Acquisition rate
30 Hz max

Power Requirements
+5V/500mA (typ.)
±12V/100 mA (typ.)

Operating Temperature
0°C to +70°C

I/O Address Requirements
4 I/O Bytes

Ordering Guide

Model PC-73 . . . . **$**
Precision thermocouple system: PC plug-in board, auxiliary screw-terminal board, connecting cable and software.

Figure 8.18 Precision Thermocouple Input Board *(Courtesy Industrial Computer Source)*

computer. Select an input channel other than the one used in the earlier assignment. Place a voltmeter across the fixed 10 k ohm resistor and observe the rise and fall of the voltmeter, as well as the digital value on the computer screen.

Prepare a table containing at least 10 readings showing a comparison between the measured voltage value and the computed value as indicated on the computer screen. Provide a comparison between the voltmeter readings and the voltage calculated by the computer. Provide a statement of percent error between the two readings. Demonstrate the operation of this circuit to your instructor.

Problem #3: Develop the necessary circuitry so that thermistor shown in Figure 8.1, together with the PIO-12 I/O board, will control a 120-volt lamp.

Write the necessary program so that when the output of the thermistor circuit (V_o) is equal to 3.5 volts the computer will turn ON the 120 volt light bulb. The bulb should remain lit as long as the analog voltage is $\geq$ 3.5 volts. When the analog input falls below 3.5 volts the lamp should be turned OFF.

Using an appropriate heat source and cooling method, cause the temperature of the thermistor to rise and fall. Observe the resulting operation of the 120 volt light bulb. **Use caution when wiring and handling the 120 volt circuit**.

Demonstrate the operation of this circuit to your instructor.

Problem #4: Modify the software for Problem #3 so that the circuit will behave in the opposite way. As the temperature of the thermistor increased and the voltage V_o voltage rises to 3.5 volts the light bulb should turn OFF and a 120 volt fan should be turned ON to provide cooling of the thermistor. Cooling the thermistor should continue until the resulting voltage generated by the thermistor circuit falls below 3.4 volts. At this point the computer should cause the light bulb to turn ON, beginning the heating process. The span or "window" of 3.5 to 3.4 volts will provide a range of temperature that the computer will attempt to maintain, while alternately heating and cooling the thermistor as needed to maintain the temperature.

Hold or mount the thermistor directly above the light bulb and locate the fan in the same proximity. Cause the program to execute and observe the resulting "controlled process." With careful placement of the bulb, thermistor, and fan, the computer should maintain the temperature in the vicinity of the thermistor within the range defined by the 3.5 to 3.4 range.

This is not a trivial assignment—some tinkering with the placement of the fan, thermistor, and lamp may be necessary so that the closed loop control will be effective. Also the digital "set points" that correspond to the 3.4 and 3.5 volt values will need to be determined and incorporated into the program. Demonstrate the operation of this circuit to your instructor.

Problem #5: Rewrite either of the above assignments so that the software uses the DAS-4 Assembly Language Subroutines provide by Keithley MetraByte Corporation. The program may by written in BASIC or C and must use the CALL function. Demonstrate the operation of this circuit to your instructor.

CHAPTER 9

Interfacing to the PC Bus: Analog Output

DIGITAL-TO-ANALOG CONVERTERS

Chapter 8 focused on the operation and application of analog-to-digital converter boards and their important role in interfacing analog input signals to a computer. This chapter reverses the signal conversion process and takes a close look at *digital-to-analog* (D/A) *converters*. As was pointed out in Chapter 8, input signals created by sensors in industrial and scientific environments are often analog electrical signals. Most physical phenomena measured or monitored are continuously variable over a specific range and do not change in discrete steps. Pressure, temperature, and speed are all analog variables that are found in the real world and are of interest to users of DA & C systems. Just as there are digital I/O boards that provide digital output signals, there are analog I/O boards that provide analog output signals.

Figure 9.1 shows a typical closed-loop analog process. In this system, the process produces a measurable analog value that is changed by a transducer from a physical quantity to a varying electrical signal. A signal conditioning circuit, either external or internal to the A/D conversion board, prepares this signal for the converter. The converter converts the analog signal to a digital form that can be used by the computer. After the computer accepts the converted data, and the operating software program has reacted to the incoming information, the computer outputs the necessary changes to the system through the digital-to-analog (D/A) converter.

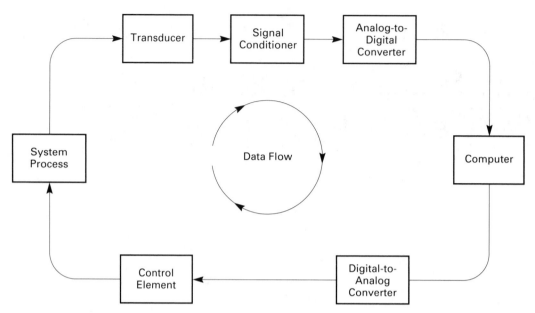

Figure 9.1 Closed Loop Analog Process

The analog output signal is brought to a control element that is part of the process being controlled. The control element alters the system process as called for by the software. The control element is often a valve, variable speed motor, or other device that can be proportionally controlled by either a varying voltage signal or a varying 4–20 mA current. Typically, the process of inputting the analog signal, performing a computation and analysis sequence, and outputting the analog control signal is repeated continuously. Thus the controlled system is constantly monitored and uses the principle of closed loop control to keep the process system operating within predetermined limits.

Typical of a process that involves both the input and output analog signals is a systems such as a computer-controlled industrial gas furnace. The input analog signal is created by a thermocouple which measures the heat produced by a gas, coal, or oil-fired burner. This signal is fed to an A/D board for conversion before being read by the software in the controlling computer. Depending upon the intent of the software program and the nature of the process being monitored, the fuel supply can be increased or decreased by proportionally opening or closing the fuel supply valve.

Figure 9.2 illustrates the use of both input and output analog signals in a closed loop control process. Digital-to-analog converters are valuable tools as they allow computers to control analog processes through providing analog signals to proportional devices or other similar analog devices. The analog output from a computer can also be fed to a loud speaker so that audio tones can be reproduced from digital data.

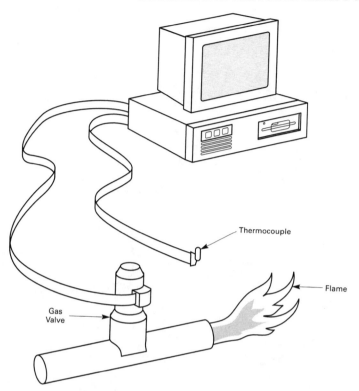

Figure 9.2 Gas Valve Control Example

Computers can be used to monitor processes that occur very rapidly, recording the measurements for later analysis. During this analysis the recorded data is converted from digital form to analog form and can be used to drive an oscilloscope display to reproduce the original analog event. A classic example of an application of D/A converter is the popular compact disk recorder found in many homes and automobiles. The operation of this device is based on the storage of music in digital form on the *compact disk* (CD). When this data is retrieved, the CD player converts the digital signals to an analog form before the voice or music signal is sent to the loud speakers.

Figure 9.3 is a block diagram of a simple D/A converter. The digital data is transferred to the converter through the data lines, and an analog voltage or current output is produced that is proportionate to the digital input.

A four-input D/A converter based on a summing operational amplifier is shown in Figure 9.3. As with any summing amp, the voltages at the input bits, A through D, are added and then applied to the op amp. Notice that the feedback resistor, R_F, is 1 K ohm. This feedback resistor in combination with the input resistors determine the value or "weight" of each of the digital inputs. The output signal from the converter is the sum of the input voltages multiplied by the ratio

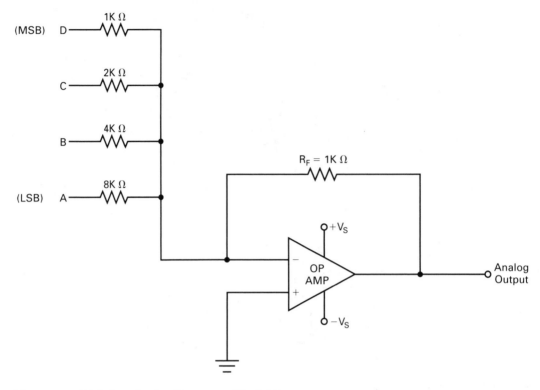

Figure 9.3 Digital-to-Analog Converter Block Diagram

of the feedback resistor to the corresponding input resistor. This voltage and resistance relationship can be expressed as:

$$V_{OUT} = -(V_D + \frac{1}{2}V_C + \frac{1}{4}V_B + \frac{1}{8}V_A)$$

For example, if the digital input were 1010, bits D and B would be a TTL high (+5 volts) and bits A and C would be a TTL low (0 volts). The resulting analog out would be calculated as:

$$V_{OUT} = -(5 + 0 + \frac{1}{4} * 5 + 0)$$

$$= -6.25 \text{ V}$$

Although the idea presented in Figure 9.3 works in theory, this approach has limitations in actual practice. The difficulty with this circuit involves the number of inputs needed for the D/A converter. As the number of digital inputs is increased, the size of the input resistors must also increase. A 12-bit converter, for example, would require that the *most significant bit* (MSB) resistor be a 1 K ohm while the *least significant bit* (LSB) resistor would be 2 Meg ohm. This variation in resistance presents problems during the manufacturing process. It is difficult to produce integrated circuits with such widely varying resistor values.

To overcome this problem, a different approach must be taken to produce a more flexible and accurate digital-to-analog converter. Figure 9.4 shows the basic construction of a *R/2R ladder digital-to-analog converter*.

In the R/2R type of converter, the digital inputs control switches that apply a reference voltage to a single voltage divider. The digital bit determines which portions of the circuit are used. The MSB applies the reference voltage through one resistor to the input of the op amp. The next lower bit has the same size input resistor as well as series resistor that is one-half that size. Each subsequent input must use this series resistor as well as its own resistor. The effect of these cumulative resistances is to allow the weighing of the digital inputs while keeping the resistor values relatively close to a single common value. In this example, the 2R resistors might have a value of 2 K ohms while the R resistors would have a value of 1 K ohm. This type of D/A converter is typical of those used in commercial devices.

Two characteristics of an A/D converter that are important to the circuit designer, engineer, and technician are step size and resolution. *Step size* is the total output voltage range of the converter divided by the number of binary steps that

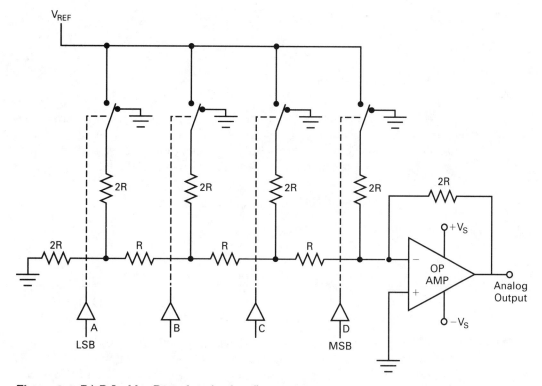

Figure 9.4 R/2R Ladder Digital-to-Analog Converter

the converter is capable of reacting to. In an 8-bit converter, there are 255 steps from minimum to maximum. If a D/A converter has an output range of 0 to -5 volts, the step size would be:

$$\text{Step Size} = \frac{V_{max} - V_{min}}{\# \text{ of steps}}$$

$$= \frac{0 - (-5) \text{ volts}}{255}$$

$$= 0.0196 \text{ volts per step}$$

This value of step size means that for each increment in the digital input, there would be a 0.0196 volt change in the analog output from the converter.

Resolution can be expressed as a voltage value or as a percentage of the maximum output voltage. Expressed as a voltage, resolution is equal to the step size. In the above example, an 8-bit converter with a range of 0 to -5 volts, the resolution is 0.0196 volts per step or per bit. When expressed as a percentage, resolution is the ratio of the step size to the total output range. Again, in the above example this resolution could be calculated as:

$$\text{Resolution} = \frac{\text{Step Size}}{\text{Total Range}}$$

$$= \frac{0.0196}{5 \text{ volts}}$$

$$= 0.00392 \text{ or } 0.392\%$$

Although percent of resolution was calculated by comparing step size to total output range, this percentage is ultimately determined by the number of digital input bits that control the converter. All 8-bit converters have the same percentage of resolution no matter the range of their output. Consider the following example:

$$\text{Resolution} = \frac{1}{2^n - 1}$$

$$= \frac{1}{2^8 - 1}$$

$$= \frac{1}{255}$$

$$= 0.00392 \text{ or } 0.392\%$$

A 10-bit converter with the same voltage limits of 0 to -5 volts has a resolution of 0.0049 volts per step. This equates to a percentage of resolution of 0.0977%. The advantage of a 10-bit converter becomes evident when the resolution between an 8-bit and a 10-bit converter is compared. The 10-bit converter has a resolution that is four times better than an 8-bit converter. This is a significant increase in resolution, and it may be an important factor when choosing a particular D/A converter.

THE DAC-02 DIGITAL-TO-ANALOG CONVERTER

Figure 9.5 describes one commercially available D/A converter, identified as the model DAC-02 produced by Keithley MetraByte Corporation. This converter is described as a 12-bit, dual-channel digital-to-analog converter. The designation "dual-channel" means that there are two separate converters on the board. Each converter is capable of being configured in one of two output modes and each channel is capable of being set to produce different output voltages. The analog output signal can be either a unipolar or a bipolar voltage or a varying 4–20 mA current.

Typical output ranges for this board, and similar analog output boards, are 0 to +5 volts, 0 to +10 volts, 0 to −5 volts, −10 volts to +10 volts, or 4–20 mA current loop. This flexibility of output voltage or current makes the DAC-02 appropriate for a wide variety of applications where analog voltage or current loop output is desirable. Figure 9.6 shows a block diagram of the DAC-02. The two output channels and the two reference voltages are clearly shown in the figure.

Like other I/O boards, the DAC-02 is a half-slot sized card. Its I/O port address is set by a 8-position DIP switch similar to other Keithley MetraByte products. Selecting the board address and setting the DIP switch follows the same procedure and precautions as other I/O boards. Refer back a few chapters to Figure 8.5. This figure can be used as a reference when setting the address switches. No interrupt capability is provided since this board is strictly an output device. Signal output from the board is through a DB25 connector on the rear plane of the board. All of the output pins as well as two voltage references are available at this connector. Figure 9.7 shows the pin connections for the DAC-02.

Before using the DAC-02 the output range for each channel must be selected. This board is capable of several types of output and a decision regarding the output range for each of the two output channels is necessary.

Minimum		*Maximum*
0	to	−5 v
0 v	to	+5 v
−10 v	to	+10 v
0	to	+10 v
4 mA	to	20 mA

The output range is determined by voltage references on the board or by external references provided by the user. The DAC-02 provides both a −5 volt and a −10 volt reference voltage on the board. The −5 volt reference (−5VREF) provides a reference voltage that is used with either the bipolar or unipolar output as long as the output voltage is rated at a maximum of 5 volts. The −5 volt reference

2-CHANNEL, 12-BIT ANALOG OUTPUT BOARD

DAC-02

FEATURES

- 2 analog output channels
- 12-bit resolution
- +5, +10, +5, +10 V output ranges
- 4 – 20 mA current loop capability
- Plugs directly into the IBM PC
- Calibration software included

APPLICATIONS

- Servo control
- Programmable amplifier
- 12-bit resolution voltage source
- Function generator

Since data is represented in 12 bits, it is written to each D/A in 2 consecutive bytes. The first byte contains the 4 least significant bits of data. The second byte contains the most significant 8 bits of data. The least significant byte is written first and is stored in an intermediate register in the D/A (having no effect on the output). When the most significant byte is written, its data is combined with the stored least significant data and presented to the D/A converter, thus assuring a single-step update. This process is known as double buffering.

The DAC-02 is a 5-inch-long "half-slot" board suitable for use in IBM PC/XT/AT and all compatibles. The DAC-02 is addressed as an I/O device using eight I/O locations and may have its I/O address set to any 8-bit boundary in the 255 – 1023 (decimal) I/O address space. The board uses the internal +5 V, +12 V and –12 V computer supplies and consumes 850 milliwatts of power.

FUNCTIONAL DESCRIPTION

The DAC-02 provides two independent double-buffered, 12-bit multiplying D/A channels plus interface circuitry. The D/A converters can be used with a fixed DC reference as conventional D/A. On board references of –5 V and –10 V provide output ranges of 0 – 5 V, 0 – 10 V, –5 V and ±10 V and 4 – 20 mA for process control current loops. Alternatively, the D/A may be operated with a variable or AC reference signal as multiplying D/A; where the output is the product of reference and digital inputs. With an AC reference, the unipolar outputs provide 2-quadrant multiplication and the bipolar outputs provide 4-quadrant operation. Twelve-bit accuracy is maintained up to 1 kHz.

SOFTWARE

Software is included with the board and contains a calibration program for adjusting the various ranges on the board. The software presents a picture of the board and points to the appropriate potentiometer for adjustment. An installation program illustrates the base address switch settings based on the Decimal of "Hex" value entered. Finally, a fully-documented program that describes in detail the simple steps needed to program the DAC-02 is included. The board uses the OUT command in BASIC.

Figure 9.5 DAC-02 Digital-to-Analog Converter Board *(Courtesy Keithley MetraByte Corporation)*

is also used for the 4–20 mA current output. If the user wants an output with a maximum of +5 volts, no matter the lower limit, the output range select jumper must be set to −5 volts. Since the upper limit of the output is +5 volts, regardless of whether the minimum output is 0 volts in the case of unipolar output or −5 volts in the case of bipolar output, the reference voltage must be configured for −5 volts.

If the maximum output voltage range is to be +10 volts, no matter the minimum voltage value, the reference must be −10 volts. Other voltage references

BLOCK DIAGRAM

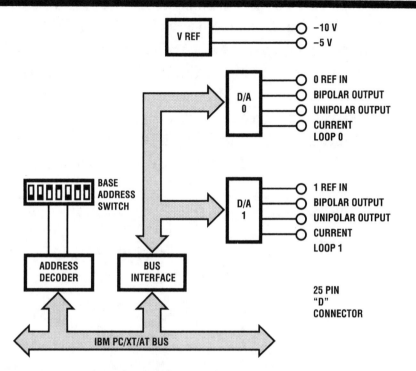

Figure 9.6 DAC-02 Block Diagram *(Courtesy Keithley MetraByte Corporation)*

may be chosen by supplying the reference from an external source. Consult the DAC-02 User Manual for specific information on supplying an external reference voltage.

Once the range has been set, the user can select either the unipolar or bipolar output by wiring the corresponding pins on the DAC-02's output connector. Figure 9.7 shows the pinout for that connector. Refer to Figure 9.7 and observe that the two channels are labeled *D/A #0* and *D/A #1*. Each channel has separate output pins devoted to bipolar output, unipolar output, and 4–20 mA output. Ground pins are also provided on this connector. If the desired output from channel #0 is to be 0 to +5 volts (unipolar) the appropriate jumper connection must be made to provide the −5 volt reference to channel #0 input (wire pin 21 to pin 22). After the jumper connection is completed, the unipolar analog output is available at pin 24 on the DB25 connector. The following couple of digital values and corresponding analog outputs are shown as examples of the range of analog output available from this board.

CONNECTOR PIN ASSIGNMENTS

A rear view of the 25-pin D connector is shown below. The DAC-02 board has a female DB25 socket and a DB25P solder cup plug is required to make connections (Keithley MetraByte part # SMC-25). Usually only 3 or 4 wires (D/A outputs and ground) will be required for connections, so that a multi-wire flat cable is not required. (Note: 25-pin D connectors are identical to RS-232C connectors). Output range selection is controlled by jumpering pins on the I/O connector or on the STA-U screw terminal board.

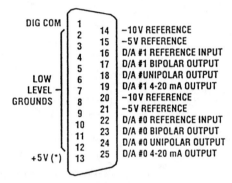

* 5V POWER FROM THE COMPUTER IS SUPPLIED ON
PIN 13. IF YOU USE THIS POWER AVOID SHORTING OR
OVERLOADING OF THE COMPUTER POWER SUPPLY.

Figure 9.7 DAC-02 Pin Connections *(Courtesy Keithley MetraByte Corporation)*

D/A Value	Unipolar Output	Bipolar Output
0	0 volts	+5 volts
2048	2.5 volts	0 volts
4095	5.0 volts	−5 volts

Programming the DAC-02

Programming the DAC-02 can be done in BASIC or C. To output a specific analog voltage, the appropriate digital value must be sent to the DAC-02. The only complication to this process is that the DAC-02 is a 12-bit device and, therefore, the user must send two bytes to the D/A board. The following internal registers and their addresses are used to hold the data bits as they are converted from digital binary bits to an analog signal.

Address	*Purpose*
Base + 0	Channel 0 least significant byte
Base + 1	Channel 0 most significant byte + update
Base + 2	Channel 1 least significant byte
Base + 3	Channel 1 most significant byte + update

During execution of the software program, the least significant byte (LSB) of the channel being used should be written first. Writing the most significant byte to the DAC-02 serves two purposes. The first role of writing the data is to complete the 12 bits that are necessary to form the 12-bit word that will be converted to a proportional analog voltage. Secondly, writing to the MSB causes the converter to UPDATE that channel output. In the context of the DAC-02, the UPDATE command function starts the conversion process. In other words, the act of writing to the register at the address base + 1 or the register at the address base + 3 causes the D/A board to do the conversion for channel 0 or channel 1 and output the resulting analog signal. The format of the two bytes used with the DAC-02 is the same as the format for the DAS-8, 12-bit Analog-to-Digital input board that was covered in Chapter 8. Figure 9.8 describes the required output format.

To calculate the two digital values that will cause the DAC-02 to output the proper analog value, the following formula should be used.

$$\text{Bipolar} \quad D_{out} = ((\text{Volts} - V_{ref}) * 2048) / (V_{ref} * -1)$$

$$\text{Unipolar} \; D_{out} = (\text{Volts} * 4095) / (V_{ref} * -1)$$

Figure 9.9 provides a programming example specifically written for the DAC-02. This program asks the computer user, through screen prompts, to enter the output voltage desired within the rage of -5 volts and $+5$ volts. The program processes this information and outputs the specified voltage on channel D/A #0. The program also determines the inverse of the specified voltage and outputs that voltage on the other channel, D/A #1. Each time the computer user specifies a new voltage within the limits of -5 to $+5$ volts the computer will output the new analog voltage on the two respective bipolar output pins.

Besides analog voltage outputs each of the D/A converters on the DAC-02 can provide 4-20 mA of current loop output. The resolution of this variable output signal is .0038 mA per bit. Figure 9.10 illustrates a current loop circuit and the connections between the circuit and the DAC-02 output connector.

The current loop power supply can range from a minimum of eight volts to a maximum of 36 volts. In typical applications, the current loop power supply is specified at 12 or 24 volts. As was explained in Chapter 3, the current loop form of circuit is ideally suited for use where the computer is a considerable distance from the controlled analog device. In this form of control circuit, the current does not degrade over distance because there is no (IR) drop in the long length of cable.

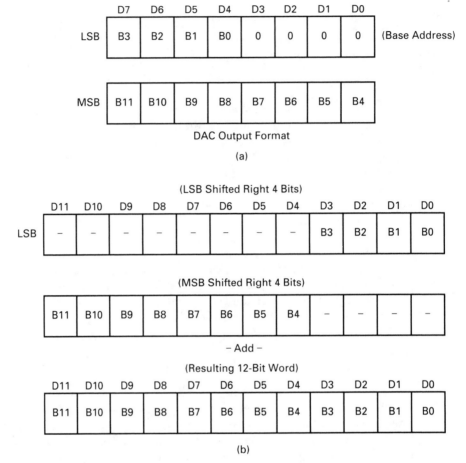

Figure 9.8 DAC Output Format

Current loop circuits are widely used in industry because of this desirable long-distance characteristic along with the circuits ability to offer a high level of noise immunity.

OTHER DIGITAL-TO-ANALOG CONVERSION PRODUCTS

Data Translation, Inc. produces a product that competes with the DAC-02. The product, identified as the model DT2815, has expanded channel capability. Figure 9.11 shows a picture of the D/A board offered by Data Translation, Inc.

The DT2815 provides eight output channels with 12-bit resolution on each channel. This board uses a single 12-bit D/A converter chip which is multiplexed to eight output channels. Each channel is equipped with a sample-and-hold circuit

```
#include <stdio.h>
#include <dos.h>              /* Needed for inp() and outp()        */
#include <conio.h>            /* Needed for gotoxy()                */
#include <ctype.h>            /* Needed for toupper                 */
#define BASE 0x300                 /* Board Base address   */
#define CHAN_0_LO BASE             /* CHANNEL 0 LO BYTE    */
#define CHAN_0_HI (BASE+1)         /* CHANNEL 0 HI BYTE    */
#define CHAN_1_LO (BASE+2)         /* CHANNEL 1 LO BYTE    */
#define CHAN_1_HI (BASE+3)         /* CHANNEL 1 HI BYTE    */

void main(void)
   {
    unsigned char ch,lo,hi;
    int temp;
    float Volts;
    do
      {
        clrscr();                          /* Clear the screen               */
        gotoxy(25,10);
        puts("Enter the desired output voltage");
        scanf("%f",&Volts);
        if((Volts>5)  ||  (Volts<-5))
           {
            gotoxy(25,20);
            puts("ERROR - Voltage out of range   -5 Volts to +5 Volts");
            gotoxy(25,21);
            puts("Press any key to continue");
            getch();
            ch='Y';
            continue;
           }
        temp=(int)((Volts + 5.0) / -0.002442);  /* Calculate digital value   */
        lo = temp & 0x0f;                        /* Isolate lowesT 4 bits     */
        lo <<= 4;                                /* Shift them into position  */
        hi=temp>>4;                              /* Shift the remaining bits  */
        outp(CHAN_0_LO,lo);                      /* output the low byte       */
        outp(CHAN_0_HI,hi);                      /* output the hi byte & convert
/* These lines find the oposite polarity voltage                              */
        temp=(int)((-Volts + 5.0) / -0.002442); /* Calculate digital value   */
        lo = temp & 0x0f;                        /* Isolate lowes 4 bits      */
        lo <<= 4;                                /* Shift them into position  */
        temp>>=4;                                /* Shift the remaining bits  */
        hi = temp;                               /*    and put them into hi   */
        outp(CHAN_1_LO,lo);                      /* output the low byte       */
        outp(CHAN_1_HI,hi);                      /* output the hi byte & convert
        gotoxy(25,20);
        puts("Do you want to do another?  Y/N");
        ch=getch();
        ch=toupper(ch);
      }
    while(ch=='Y');                    (a)
   }
```

Figure 9.9(a) DAC-02 Programming Example

to provide signal stability. Figure 9.12 shows a block diagram of the DT2815 A/D board.

Engineers and technicians developing solutions for applications that require more than a few analog output signals have two choices. One choice is to use more than one D/A board in the computer or the other choice is to install a single board

```
'
'
'
'
'
CONST BOARD = &H300
CONST CHAN.0.LO = BOARD + 1
CONST CHAN.0.HI = BOARD + 1
CONST CHAN.1.LO = BOARD + 2
CONST CHAN.1.HI = BOARD + 3

DO
  CLS
  LOCATE 10, 25
  INPUT "Enter the desired output voltage ", Volts
  IF ((Volts > 5!) OR (Volts < -5!)) THEN
      LOCATE 20, 25
      PRINT "ERROR - Voltage out of range   -5 Volts to +5 Volts"
      LOCATE 21, 25
      PRINT "Press any key to continue"
      WHILE (INKEY$ = "")
      WEND
      CH$ = "Y"
  ELSE
      TEMP = (Volts + 5!) / -.002442      ' Calculate the digital value
      LO = TEMP AND &HF                   ' Isolate the lowest 4 bits
      LO = LO * 16                        ' Shift them into position
      HI = TEMP / 16                      ' Get the highest 8 bits into hi
      OUT CHAN.0.LO, LO                   ' Output the low byte
      OUT CHAN.0.HI, HI                   ' Output the hi byte and convert
' These lines find the oposite polarity voltage
      TEMP = (-Volts + 5!) / -.002442     ' Calculate the digital value
      LO = TEMP AND &HF                   ' Isolate the lowest 4 bits
      LO = LO * 16                        ' Shift them into position
      HI = TEMP / 16                      ' Get the highest 8 bits into hi
      OUT CHAN.1.LO, LO                   ' Output the low byte
      OUT CHAN.1.HI, HI                   ' Output the hi byte and convert
      LOCATE 20, 25
      PRINT "Do you want to do another?  Y/N"
      DO
         CH$ = INKEY$
      LOOP WHILE ((CH$ <> "Y") AND (CH$ <> "y") AND (CH$ <> "N") AND (CH$ <> "n"))
  END IF
LOOP WHILE ((CH$ = "Y") OR (CH$ = "y"))
```

Figure 9.9(b) DAC-02 Programming Example

with lots of D/A channels. Using more than one I/O board can be done by installing two or more boards like the DAS-02 or the DT2815 in a computer. Multiple boards can be installed in a computer as long as care is taken to avoid over lapping addresses.

Another option is to use a board that is specifically designed to accommodate a large number of analog outputs. One board that fits this specification is the model AOB12 produced by Industrial Computer Source, Inc. The AOB12 provides 12 independent analog output channels. Each channel is equipped with a separate 12-bit D/A converter which can be configured for either unipolar or bipolar outputs

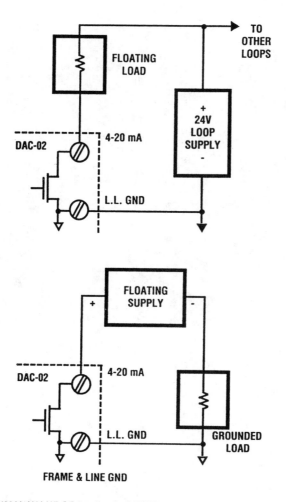

A DECIMAL VALUE OF 0 = 4 mA, A DECIMAL VALUE OF 4095 = 20 mA

Figure 9.10 4-20 mA Current Loop Application *(Courtesy Keithley MetraByte Corporation)*

and the typical range of output voltages. Figure 9.13 shows a picture of the AOB12 D/A board.

This board, though expensive, does occupy one slot in the computer and yet it provides a lot of D/A capability.

INTERFACING PROBLEMS

Problem #1: Write the necessary software so that the output of the digital-to-analog converter is a triangle waveform. The waveform should be adjusted so that the rise

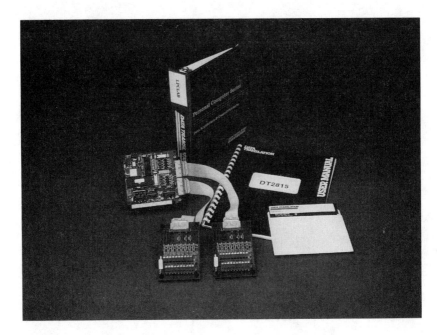

Figure 9.11 DT2815 Analog Output Board and External Screw Terminal Panels (Courtesy Data Transaction, Inc.)

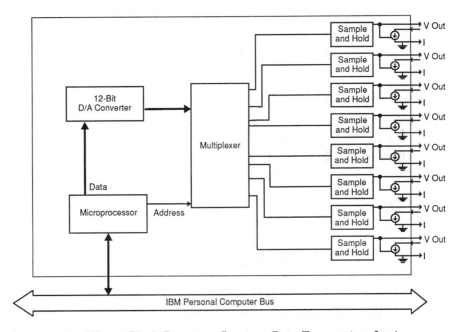

Figure 9.12 DT2815 Block Diagram (Courtesy Data Transaction, Inc.)

Model AOB12

The AOB12 is ideal for the user who needs a large amount of D/A capability. This card contains nothing but D/A.

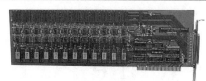

FEATURES
- **30 to 50KHz Throughput**
- **12 Independent 12-Bit DACS**
- **Parallel Triggering Available on 8 Channels**
- **Settling Time < 2 usec**
- **Unipolar & Bipolar Modes**
- **Bipolar Mode Supports Remote Sensing**
- **Software Provided With Board**

Specifications

Channels: 12
Resolution: 12-bit (1/4096)
Ranges
  -5 to +5, -10 to +10 (8 channels)
  0 to +5, 0 to +10 (12 channels)
Linearity: ±LSB

Settling Time
  <2 u sec, -5 to +5 step

I/O Address Requirement
  24 I/O Bytes
Connector: 37-pin, D-shell, female

Ordering Guide

Model AOB12 .... $
Model UTB-K.. .... $
  Termination Board

Model C1900 .... $
18" (45.72 cm) Cable

Figure 9.13 Model AOB12 D/A Multi-channel Converter Board
(Courtesy Industrial Computer Source, Inc.)

and fall of the voltage output can be monitored by an oscilloscope. Connect the output waveform to an oscilloscope, and show the triangle waveform to your instructor.

Problem #2: Modify the software developed for Problem #1 so that the output waveform is a sawtooth. Demonstrate the waveform to your instructor.

Problem #3: This assignment investigates whether the DAS-4 and DAC-02 can be combined to form a computer controlled amplifier. Connect the input of the DAS-4, A/D converter, to an analog source such as a sine wave signal generator. Connect an oscilloscope to the output of the DAC-02 digital-to-analog board. Write the necessary software so that the analog signal input to the DAS-4 is digitized by the A/D board, amplified within the computer by a factor of 2, and reconverted to an analog waveform by the DAC-02 A/D board. Demonstrate the operation of this circuit to your instructor.

Problem #4: Connect two different variable voltage signal sources to the DAS-4. One signal source should be connected to channel 0 and the other signal should be connected to channel 1. Write the software program so that the output of the DAC-02 is an error voltage that represents the difference between the two analog input signals. Demonstrate the operation of this circuit to your instructor.

Problem #5: Connect the output of the D/A converter to an appropriate transistor driver and a small incandescent lamp. Write a program so that the computer operator can slowly raise and lower the brilliance of the lamp. Demonstrate the operation of this circuit to your instructor.

CHAPTER 10

Interfacing to the PC Bus: Serial I/O

Communication is important whether it be between two or more people or between parts of a computer control system. Communication techniques that have been presented in this text up to this point have been in a parallel data format. That is, eight bits of information has been transferred at once. This was the format used with the input and output software commands when using the PIO-12 digital I/O board as presented in Chapters 5 and 6. During the introduction to digital communications presented in Chapter 2, it was pointed out that parallel data transfer is faster than serial data transfer. While parallel data transfer is faster the distance that digital data can be transmitted is a limiting factor. Chapter 10 will investigate serial data transfer focusing on topics including data transfer protocol, wiring connections, and serial specifications.

Figure 10.1 is a block diagram of a typical PC serial port. This port is often an add-on board similar in appearance to the PIO-12 digital I/O board or the DAS-4 A/D board. In the add-on form, the serial board is plugged into the bus just as other optional boards are inserted into the computer.

In recent years, most computer manufacturers recognize the importance and popularity of serial communications and are incorporating the serial function directly into the computers system (main) board. The *main board*, often called a *mother board* or *system board* will have the CPU, memory, and functions such as a serial port included in the main system board. Whether the serial port is an add-on inserted into the system bus or located on the on-system board, the computer

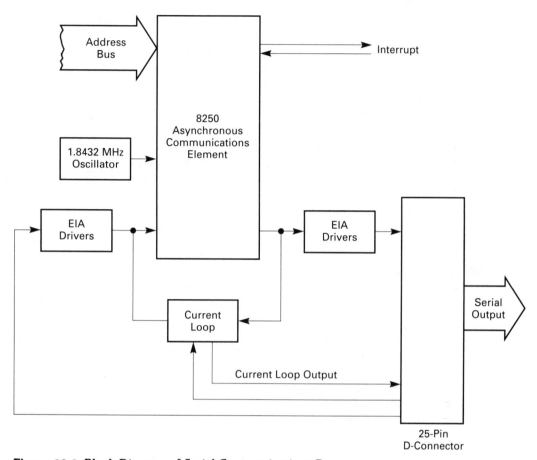

Figure 10.1 Block Diagram of Serial Communications Port

sends data to external serial devices through the serial port. Serial data, by definition, is transmitted one bit at a time. The transmitted data is unique in that it is not TTL compatible. Refer to Chapter 3 for a review of RS-232 serial signal levels.

HARDWARE HANDSHAKING

The data that is to be transmitted through the serial port is presented to the serial board by the microprocessor in the parallel format under which the processor operates. Circuitry on the serial board, or within the serial function on the mother board, performs parallel-to-serial conversion on the data. Similarly, data received from external serial devices goes through the reverse process. Data entering the computer from an external serial device is converted from serial-to-parallel. This incoming data is placed on the computers parallel oriented bus system. The parallel data is ready to read by the microprocessor, stored in memory, or displayed on the CRT.

Figure 10.2 illustrates two handshaking examples. *Handshaking* gives serial ports the option of communicating between ports for the purpose of controlling the flow of data bits between the transmitting and receiving ports.

These examples range from the simplest type of handshaking, shown in Figure 10.2(a), to a more complex example of handshaking shown in Figure 10.2(b). The specific requirements and needs of the serial communications determines what type of handshaking will be used. In Figure 10.2(a) the computer transmits the serial bits without regard for whether the receiving device is ready to receive the information. This form of serial communications is used when the computer and the external device, possibly another computer, or other receiving device operate at a much faster rate than the speed at which the data is transmitted through the serial port. Often the limiting element in the communications link is the serial port.

Figure 10.2(b) shows a more sophisticated serial connection. In this example four handshaking lines are used besides the *data transmit* (TxD) and *data receive* (RxD) lines. The *data terminal ready* (DTR) output from one device, or computer, is connected to the *data set ready* (DSR) input to the other device. The DSR pin is coupled back to the *data carrier detect* (DCD) *pin*. This connection guarantees that the handshaking line will be correctly driven by the DTR on the receiving end of the cable. The DSR and the DCD lines are used in the communications process by one of the devices to inform the other device that it is connected to the serial link and working. The *request to send* (RTS) line from each device is connected to the *clear to send* (CTS) line of the other device. This connection controls when data is transferred between the two connected devices. The transmitting device activates its RTS line when it desires to begin sending data.

When the receiving device senses that its CTS input is activated it knows that there is data waiting to be sent by the transmitting device to the receiving device. Transmission would not begin, though, until the receiving device forces its own RTS line active. The RTS bit is connected to the sender's CTS line. An active signal on this line indicates to the transmitting device that the receiver is ready to receive the data. The receiving device can stop the transmission process at any time by forcing the RTS line inactive. As the RTS line goes inactive the transmitting device would detect this change suspend transmission until the RTS line again goes active.

Handshaking is useful when the transmitting device identified as *data terminal equipment* (DTE) operates much faster than the receiving device identified as *data communications equipment* (DCE). It is essential that the DCE equipment have a method of controlling the flow of data. Otherwise the data will be lost in transmission. Use of the handshaking lines involves communications software that monitors the handshaking lines responds by controlling the transmitting device (DTE).

The designations data terminal equipment (DTE) and data communication equipment (DCE) are terms that identified equipment that was in use when the RS-232 standard was established in the early 1960s.

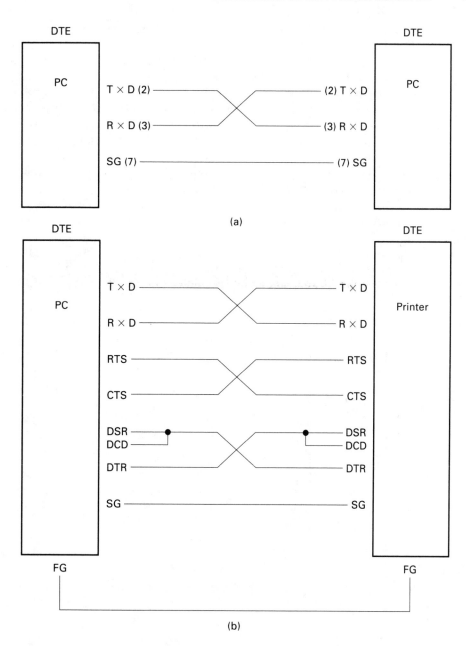

Figure 10.2 Serial Communications Handshaking Examples

The term "data terminal equipment" (DTE) was originally meant to apply to a CRT terminal or other input device. Today, the DTE designation is typically applied to a personal computer. Figure 10.2 shows two PCs connected as DTE devices. In that figure, the two DTE devices are *cross-coupled* in what is called a *null modem* configuration. A *crossover*, called a *null modem modification*, is required in the wiring between the two computers. The crossover is needed so that the data transmitted on the TxD pin from the transmitting computer is connected to the RxD data pin on the receiving computer. Because either computer can serve as the transmitting device this *crossover* must be used with TxD pins and RxD pins on both computers. Serial cables should be clearly marked or otherwise identified indicating whether they are *straight* or *cross coupled*.

The designation "data communication equipment" (DCE) originally applied to a modem or other types of data communications equipment. A modem converts the digital RS-232 signal into an analog signal for transmission over traditional telephone lines. Today, the term DCE is still applicable and is typically used to identify a modem or other similar device. Figure 10.3 shows a PC connected to a modem or other DCE piece of equipment.

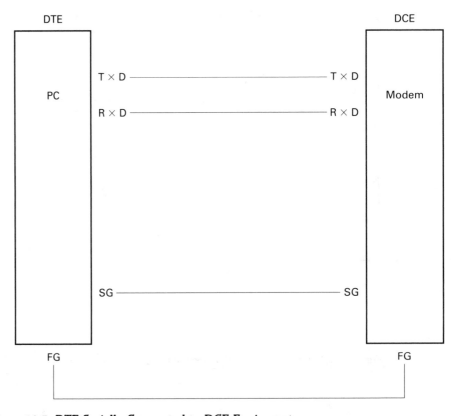

Figure 10.3 DTE Serially Connected to DCE Equipment

The wiring connection between the DTE and the DCE device is a *straight-through connection*. Pin number 2 (TxD) on the computer end of the cable is connected to pin number 2 (TxD) on the modem end of the cable. The same straight-through connections are used on all other wires connected between the DTE and DCE. The pins that are used, other than transmit data (TxD) and receive data (RxD), serve handshaking functions. Typical handshaking pins are *ready to send* (RTS), *clear to send* (CTS), *data terminal ready* (RTD), and *data set ready* (DSR). These pins inform the DCE that the DTE is ready to send information; conversely, they inform the DTE whether the DCE is ready to accept the data.

Figure 10.4 illustrates a very practical situation where an RS-232 serial cable is connecting a PC to a robot. Both the robot and the PC are identified as DTE devices. This identification is typical because most robots have a serial data port and are configured as DTEs. The cable has *signal ground* (SG) connected at both ends. Running a signal ground and preferably a *frame ground* (FG) between the two devices is important.

In some cable runnings, only three wires between the two DTE devices may be practical. This form of connection was shown in Figure 10.2(a) as an example of a simple RS-232 serial connection. In that figure only the transmit, receive, and signal ground wires were run between the two DTE devices. Handshaking wires were not used.

In another form of serial connection, the handshaking is "faked-out" by the loop-back connections that are shown in Figure 10.4. The folded-back handshaking connections allow the two DTEs to communicate serially. Each device thinks that it is conducting handshaking with the other device, when actually the handshaking lines are tricked into allowing the transmission and receiving of serial data. While fewer connecting wires with this form of connection may seem attractive, this configuration does not allow for real handshaking between the two DTEs. All handshaking must be accomplished by software control. A major shortcomming in this configuration is the lack of DTR signal. Because this signal is missing the neither device knows if the other is powered up and ready to receive data. The computer does not know if the robot is ready to receive the data. Likewise, the robot does not know the status of the computer. This form of serial communication is simple, however it does lack the sophistication of a true handshaking serial connection in which all of the handshaking connections are used.

SOFTWARE HANDSHAKING

Software handshaking is another method of controlling a serial communications process. This technique involves the use of special characters to control the data flow. Figure 10.5 provides a listing of the ASCII characters and control codes that can be transmitted through a serial port.

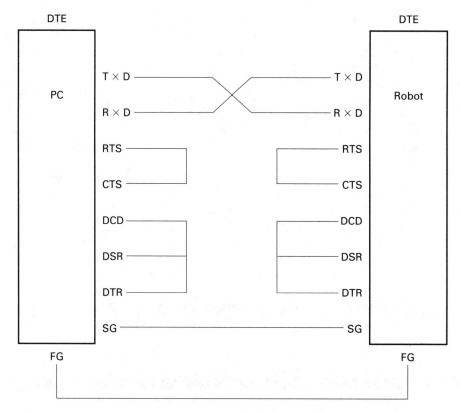

Figure 10.4 Typical Serial Connection Between Computer and Robot

The characters, letters, numbers, and punctuation are the printable codes. Other codes are used to control the data flow as well as control the external hardware. For example, the code represented by **BEL** (ASCII code 07) is a code that causes the receiver, perhaps a printer, to beep its internal "bell" or equivalent sound source. In devices not equipped with an actual bell, a tone will be output through a speaker. Another ASCII code is the Hex code 0D. This code causes a *carriage return* to occur in a printer.

Two codes can be used to control data flow. They are *data control 1* (DC1) and *data control 3* (DC3). These codes are also known as XON and XOFF, respectively. Often software is written so that before data is transmitted the receiver of the data must first send an XON code back to the sending device. If during the communications process the receiving piece of equipment wants to stop the trans-

ASCII	Decimal	Hex	Octal	Binary	Key*	ASCII	Decimal	Hex	Octal	Binary
NUL	00	00	00	000 0000	*@	SP	32	20	40	010 0000
SOH	01	01	01	000 0001	*A	!	33	21	41	010 0001
STX	02	02	02	000 0010	*B	"	34	22	42	010 0010
ETX	03	03	03	000 0011	*C	#	35	23	43	010 0011
EOT	04	04	04	000 0100	*D	$	36	24	44	010 0100
ENQ	05	05	05	000 0101	*E	%	37	25	45	010 0101
ACK	06	06	06	000 0110	*F	&	38	26	46	010 0110
BEL	07	07	07	000 0111	*G	'	39	27	47	010 0111
BS	08	08	10	000 1000	*H	(	40	28	50	010 1000
HT	09	09	11	000 1001	*I	)	41	29	51	010 1001
LF	10	0A	12	000 1010	*J	*	42	2A	52	010 1010
VT	11	0B	13	000 1011	*K	+	43	2B	53	010 1011
FF	12	0C	14	000 1100	*L	,	44	2C	54	010 1100
CR	13	0D	15	000 1101	*M	-	45	2D	55	010 1101
SO	14	0E	16	000 1110	*N	.	46	2E	56	010 1110
SI	15	0F	17	000 1111	*O	/	47	2F	57	010 1111
DLE	16	10	20	001 0000	*P	0	48	30	60	011 0000
DC1	17	11	21	001 0001	*Q	1	49	31	61	011 0001
DC2	18	12	22	001 0010	*R	2	50	32	62	011 0010
DC3	19	13	23	001 0011	*S	3	51	33	63	011 0011
DC4	20	14	24	001 0100	*T	4	52	34	64	011 0100
NAK	21	15	25	001 0101	*U	5	53	35	65	011 0101
SYN	22	16	26	001 0110	*V	6	54	36	66	011 0110
ETB	23	17	27	001 0111	*W	7	55	37	67	011 0111
CAN	24	18	30	001 1000	*X	8	56	38	70	011 1000
EM	25	19	31	001 1001	*Y	9	57	39	71	011 1001
SUB	26	1A	32	001 1010	*Z	:	58	3A	72	011 1010
ESC	27	1B	33	001 1011	*[	;	59	3B	73	011 1011
FS	28	1C	34	001 1100	*\	<	60	3C	74	011 1100
GS	29	1D	35	001 1101	*]	=	61	3D	75	011 1101
RS	30	1E	36	001 1110	-	>	62	3E	76	011 1110
US	31	1F	37	001 1111	*-	?	63	3F	77	011 1111

*The * is the CTRL (control) key pressed simultaneously with the other indicated key.

ASCII	Decimal	Hex	Octal	Binary	ASCII	Decimal	Hex	Octal	Binary
@	64	40	100	100 0000	'	96	60	140	110 0000
A	65	41	101	100 0001	a	97	61	141	110 0001
B	66	42	102	100 0010	b	98	62	142	110 0010
C	67	43	103	100 0011	c	99	63	143	110 0011
D	68	44	104	100 0100	d	100	64	144	110 0100
E	69	45	105	100 0101	e	101	65	145	110 0101
F	70	46	106	100 0110	f	102	66	146	110 0110
G	71	47	107	100 0111	g	103	67	147	110 0111
H	72	48	110	100 1000	h	104	68	150	110 1000
I	73	49	111	100 1001	i	105	69	151	110 1001
J	74	4A	112	100 1010	j	106	6A	152	110 1010
K	75	4B	113	100 1011	k	107	6B	153	110 1011
L	76	4C	114	100 1100	l	108	6C	154	110 1100
M	77	4D	115	100 1101	m	109	6D	155	110 1101
N	78	4E	116	100 1110	n	110	6E	156	110 1110
O	79	4F	117	100 1111	o	111	6F	157	110 1111
P	80	50	120	101 0000	p	112	70	160	111 0000
Q	81	51	121	101 0001	q	113	71	161	111 0001
R	82	52	122	101 0010	r	114	72	162	111 0010
S	83	53	123	101 0011	s	115	73	163	111 0011
T	84	54	124	101 0100	t	116	74	164	111 0100
U	85	55	125	101 0101	u	117	75	165	111 0101
V	86	56	126	101 0110	v	118	76	166	111 0110
W	87	57	127	101 0111	w	119	77	167	111 0111
X	88	58	130	101 1000	x	120	78	170	111 1000
Y	89	59	131	101 1001	y	121	79	171	111 1001
Z	90	5A	132	101 1010	z	122	7A	172	111 1010
[	91	5B	133	101 1011	{	123	7B	173	111 1011
\	92	5C	134	101 1100			124	7C	174	111 1100
]	93	5D	135	101 1101	}	125	7D	175	111 1101
ñ	94	5E	136	101 1110	~	126	7E	176	111 1110
-	95	5F	137	101 1111	DEL	127	7F	177	111 1111

Figure 10.5 ASCII Characters and Control Codes

mission of data, the receiving device simply sends an XOFF code back to the sending device. The transmitting device will stop sending serial data and it will wait until it receives an XON code before resuming transmission.

The XON and XOFF codes can be sent from a computer keyboard. The key strokes ⟨CTRL⟩Q cause a XON to be transmitted and a ⟨CTRL⟩S causes an XOFF to be created. These keyboard activated control signals are frequently used to manually control the transmission of data under manual intervention.

SERIAL PORT SOFTWARE

Controlling a serial port involves setting up the hardware for the proper transmission speed (*baud rate*), data format, and controlling the handshaking lines. The following baud rates are commonly used with personal computers.

Baud Rate

110
300
1200
2400
4800
9600

Serial transmission rates are established as standards through industry convention. Along with setting the baud rate, initialization of a serial port requires that the number of data bits and stop bits be specified.

Figure 10.6 shows a diagram of a serial character. In this figure notice that every data word transmitted begins with a *start pulse*.

This pulse is a logical 0 and lasts for one bit period. The next bit following the start pulse is the least significant bit (LSB) of the data word. When all bits of the data word have been sent one stop bit or two stop bits are transmitted. These *stop* bits are typically a logic 1. Additionally, a *parity* bit, an option that serves to validate the data word, maybe inserted between the MSB and the stop bits.

The number of data bits that make up a serial transmitted word is usually seven or eight bits and the number of stop bits is either one or two. If there is not total agreement between the transmitting device and the receiving device relative to the number of stop bits transmitted and the number of stop bits expected there may be complications with the transmitted data. For example if two stop bits are sent by the transmitting device and only one stop bit is expected by the receiving device the effect will be that the actual data transfer rate will be slightly less than expected. If only 1 stop bit is sent when 2 stop bits are expected by the receiving device some data may be lost. These two examples illustrate the importance of being certain that both the transmitting device and the receiving device are in agreement regarding the number of STOP bits to be included in the serially transmitted word.

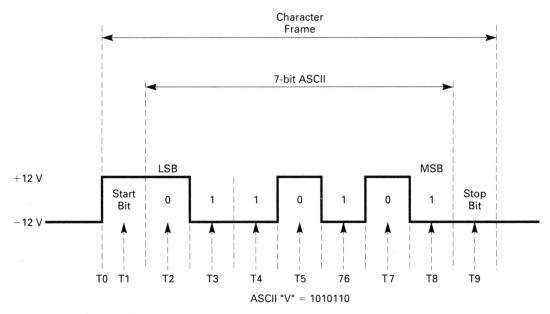

Figure 10.6 Serial Word

Parity, an optional technique used to provide error detection of transmitted data, requires the insertion of a bit into the data word. The parity bit may be a logic 1 or a logic 0 depending upon the word being transmitted and the previous-lyestablished protocol of *even parity* or *odd parity*.

Initialization of a serial port requires that the following parameters be defined before the transmission of the data; number of data bits (7 or 8), number of stop bits (1 or 2), and whether parity is used. If parity is used either odd parity or even parity must be selected.

The BASIC programming language offers a simple mechanism for transmitting data via a serial port. The command is a follows:

```
OPEN ''COM1:9600,N,8,1,BIN'' AS 2
```

This command opens serial port, COM1, as serial port and sets the baud rate to 9600 bits per second. Each serial data word would have eight data bits, one stop bit, and no parity bit. Additionally, all data transmitted will be in binary format.

The C programming language uses a different technique to set up serial ports. The technique used with the C language will be illustrated in Figure 10.17.

In real-world computer applications accuracy in data transmissions is critical. One method of assuring accuracy is through the use of interrupts. Interrupts are usually handled through assembly language routines and are beyond the scope of this text. However, the reader is encouraged to investigate assembly language programming and the use of interrupts in interfacing and process control.

SERIAL DB25 CONNECTOR

Data transmission into or out from a serial board or serial portion of the mother board is typically connected through a DB25 connector. This connector shown in Figure 10.7 has 25 pins. Notice that *transmit data* (TxD) and *receive data* (RxD) are pins 2 and 3, respectively. This convention is standard and can be expected on any DB25 serial port.

8250 SERIAL I/O

Figure 10.8 shows a block diagram of the INS8250 *asynchronous communications element* (ACE). This device is the "heart" of the asynchronous communications board. Observe that this device contains several registers. The registers are used to store all of the information necessary to configure the serial port.

These registers also hold the transmitted and received binary characters. Selection of registers is accomplished through the address and control block. Each register is assigned an address relative to the base address of the serial port. Addresses of the first two serial ports standard in most computers has been standardized. These addresses are, serial port #1 (COM1) = 3F8H, serial port #2 (COM2) = 2F8H. Figure 10.9 provides a listing of the INS8250 registers and their addresses.

HOLDING REGISTERS

As their name implies, these registers are dedicated buffers for *holding* a character that has been received but not yet read, or a character that has been sent to the serial port but not yet transmitted. The INS8250 chip is referred to as a *double-buffered ACE*. This means that there are two buffers for each function. The feature allows a second character to be sent to the serial port before the first character has been completely transmitted or read by the processor.

LINE CONTROL REGISTER

Another register in the INS8250 chip is named the *line control register* (LCR). This register holds the parameters established by the programmer that determine the format of the communications exchange. Information such as the number of data bits, the number of stop bits, and the type of parity used are held in this register. Data can be written to this to this register and read at a later time. Figure 10.10 provides insight to the bits functions of the LCR.

Bits 0 and 1. The value stored in these binary bits specifies the number of data bits in each transmitted character. The number of characters can range from five

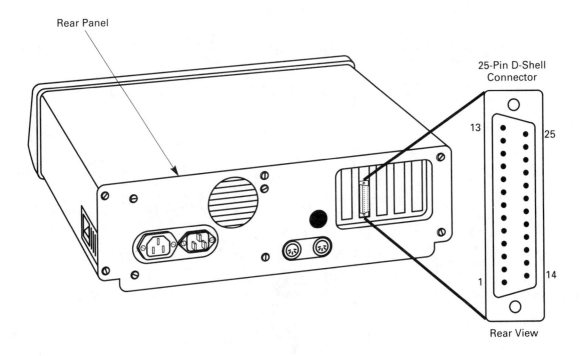

Pin	Name	Abbreviation
1	Frame ground	FG
2	Transmit data	T × D
3	Receive data	R × D
4	Request to send	RTS
5	Clear to send	CTS
6	Data set ready	DSR
7	Signal ground	SG
8	Data carrier detect	DCD
20	Data terminal ready	DTR

Figure 10.7 DB25 Serial Connector and Common Pin Definitions

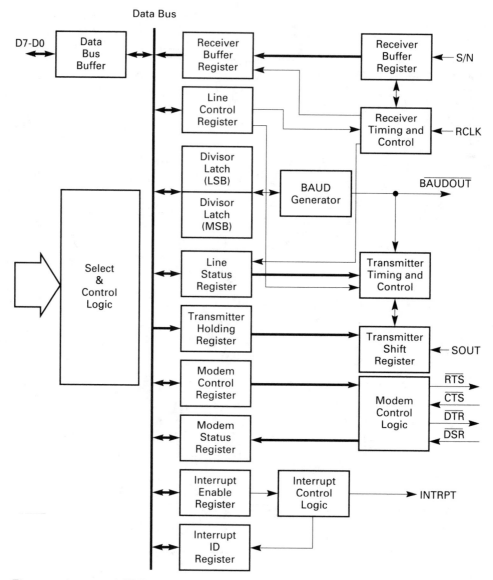

Figure 10.8 8250 ACE Registers

to eight bits. Figure 10.8(b) shows the coding scheme for determining the length of the data word.

Bit 2. This bit specifies the number of stop bits in each character string. If bit 2 is a logic 0 then 1 stop bit will be generated by the 8250 chip. If the transmitted character has six, seven, or eight data bits, and bit 2 is set to a logic 1 then two

DLAB	A2	A1	A0	Register	Address
0	0	0	0	Receiver Buffer Read). Transmitter Holding Register (Write)	3F8 (2F8)
0	0	0	1	Interrupt Enable	3F9 (2F9)
X	0	1	0	Interrupt Identification (Read Only)	3FA (2FA)
X	0	1	1	Line Control	3FB (2FB)
X	1	0	0	Modem Control	3FC (2FC)
X	1	0	1	Line Status	3FD (2FD)
X	1	1	0	Modem Control Status	3FE (2FE)
X	1	1	1	None	
1	0	0	0	Divisor Latch (Least Significant Bit)	3F8 (2F8)
1	0	0	1	Divisor Latch (Most Significant Bit)	3F9 (2F9)

Figure 10.9 8250 Register Listing

stop bits will be generated and "attached" to each transmitted word. If five data bits are chosen as the coding system for a character, then 1.5 stop bits are required to be inserted into the data word. This requirement is necessary to conform with older Baudot equipment, now obsolete, that used five bits of data.

Bit 3. This bit is identified as the *parity enable bit*. If the bit is a logic 1, a parity bit will be generated and inserted into each character string. Because parity is enabled any received characters will also be checked for a parity bit.

Bit 4. The type of parity selected, *odd* or *even*, is determined by setting a bit 4. Storing a logic 0 in this location sets the parity for odd and, conversely, storing a logic 1 in bit 4 sets the parity for even. If bit 3, parity enable, is disabled by placing a logic 0 in that location, then regardless of what bit value is in bit 4, it has no effect.

Bit 5. Stick Parity. If bit 3 and bit 5 are set to a logic 1, then when the transmitter outputs a character the local receiver detects as a logic 1.

Bit 6. This bit is identified as the *break bit*. When it is set to a logic 1 it forces SOUT (TxD) to go to the space logic level until a logic 0 is stored in bit 6. This action allows the computer to signal a terminal which is connected as part of the communications system.

Bit 7. This bit must be set to a logical 1 to access the *divisor latches*. These latches are registers that store the clock divisor that establish the baud rate of the serial communications system. Once the baud rate has been set this bit (bit 7) is reset to a logic 0.

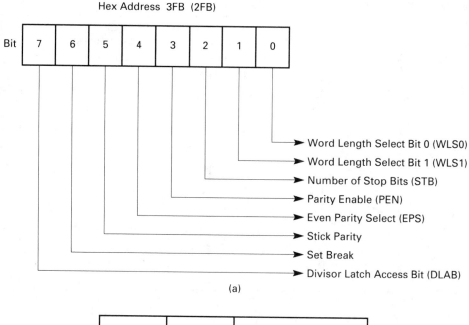

Hex Address 3FB (2FB)

Bit 7 6 5 4 3 2 1 0

→ Word Length Select Bit 0 (WLS0)
→ Word Length Select Bit 1 (WLS1)
→ Number of Stop Bits (STB)
→ Parity Enable (PEN)
→ Even Parity Select (EPS)
→ Stick Parity
→ Set Break
→ Divisor Latch Access Bit (DLAB)

(a)

Bit 1	Bit 0	Word Length
0	0	5 Bits
0	1	6 Bits
1	0	7 Bits
1	1	8 Bits

(b)

Figure 10.10 Line Control Register Bits

CONTROLLING THE BAUD RATE GENERATOR

Review Figure 10.9 and note that there are two registers for each of the base addresses and also two registers for the next higher address. This doubling of registers is necessary because when bit 7 or the LCR (the DLAB previously discussed) is reset to logic 0 these two addresses refer to the *receive buffer* and the *transmit buffer*. When the DLAB bit is set to a logic 1 these two addresses refer to the two divisor latches. The divisor latches, arranged as a LSB and MSB, are used in setting the baud rate of the communications system. The value stored in these two registers is multiplied by 16 and then used to divide the serial port clock. The serial port clock is standardized in personal computers at a frequency of 1.8432 MHz. If baud rate of 1200 baud were desired the following calculations indicate the necessary divisor value.

$$1200 = \frac{1843200}{16 \times \text{Divisor}} = \text{Divisor} = \frac{1843200}{16 \times 1200}$$

$$\text{Divisor} = 96$$

$$\text{Divisor} = 060 \text{ Hex}$$

Because the divisor latches are two bytes wide, the value (060 Hex) needs to be divided between two registers. In this example (1200 baud rate), the Hex 60 would be stored in the LSB and the value 0 would be stored in the MSB. Figure 10.11 illustrates several baud rates and the appropriate divisor values in both decimal and Hex form.

LINE STATUS REGISTER

The *line status register* (LSR) shown in Figure 10.12 is an 8-bit register that, when read, provides information to the microprocessor concerning the data transfer through the serial port.

Desired Baud Rate	*Divisor Used to Generate 16× Clock* (Decimal)	(Hex)	*Percent Error Difference Between Desired and Actual*
50	2304	900	—
75	1536	600	—
110	1047	417	0.026
134.5	857	359	0.058
150	768	300	—
300	384	180	—
600	192	0C0	—
1200	96	060	—
1800	64	040	—
2000	58	03A	0.69
2400	48	030	—
3600	32	020	—
4800	24	018	—
7200	16	010	—
9600	12	00C	—

Figure 10.11 Baud Rate Table Based on 1.8432 MHz Clock

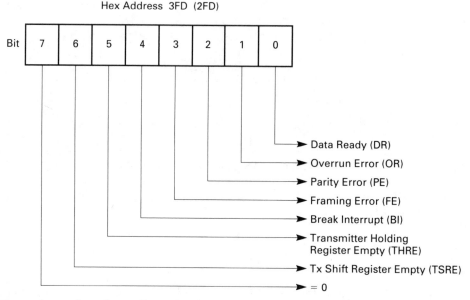

Hex Address 3FD (2FD)

Figure 10.12 Line Status Register

Bit 0. The *data received* (DR) *bit*, bit 0, is set to a logic 1 when data has been received and is ready to be read by the processor.

Bit 1. A logical 1 in this bit means that the previous receive character was lost because it was not read before a new character was received. The new character essentially overran the previous character.

Bit 2. A logic 1 in the parity error bit means that the received character has the wrong parity. When the line status register (LSR) is read, this bit is reset to logic 0.

Bit 3. If the received character did not have a valid stop bit, bit 3 in the LSR is set to a logic 1.

Bit 4. This bit is identified as the *break interrupt bit*. This bit is automatically set to a logic 1 when the received data word has been held at a space level for the length of a data word.

Bit 5. This bit is identified as the *transmit holding register empty* (THRE) *bit*. This bit signifies that the serial port is ready to accept another character to be transmitted.

Bit 6. This bit location is a read-only bit. When it is at a logic 1 level the transmitter is at idle.

Bit 7. This bit is not used and is always at a logic 0 level.

INTERRUPT REGISTERS

The 8250 chip has a broad range of interrupt capabilities. Two registers are used to control and identify sources of the interrupt. The first of these two registers is the *interrupt enable register* (IER). If the interrupt capability of the chip has been enabled and an interrupt occurs, then the interrupt output bit from the 8250 assumes a logic 1 level. This signal is connected to the hardware interrupt bus of the computer. The logic 1 on this bus signals the processor that the serial port needs attention. Figure 10.13 illustrates a bit map for the IER.

Bit 0. Each time a character is received an interrupt is generated. This bit is reset after the character has been read by the microprocessor.

Bit 1. If this bit is set to logic 1 an empty transmit holding register would cause in interrupt to occur.

Bit 2. This bit enables a change in the receiver line status of cause an interrupt.

Bit 3. This bit enables a change in the modem status to interrupt the processor.

Bits 4–7. These bits are permanently set to logic 0.

If an interrupt occurs the program software must poll a register to determine what type of event is causing the interrupt. The *interrupt identification register* (IIR) contains the code that identifies which condition (interrupt) is requesting attention.

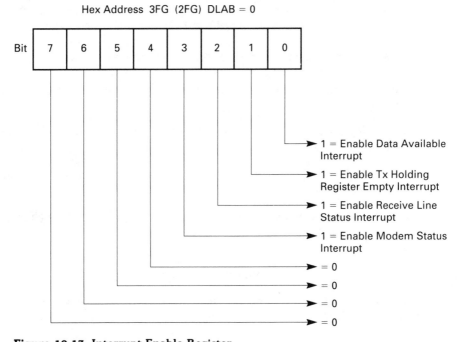

Hex Address 3FG (2FG) DLAB = 0

Figure 10.13 Interrupt Enable Register

Figure 10.14(a) illustrates the IIR register. Figure 10.14(b) lists the ID codes and the corresponding interrupt conditions. In an effort to keep software overhead to a minimum the interrupts are prioritized. Prioritizing means that some interrupts are "more important" than other interrupts.

Figure 10.14(b) lists the priority levels for each interrupt. The interrupt reset column lists what action is needed to reset the latched interrupt.

MODEM REGISTERS

The last two registers in the 8250 are used to monitor as well as control the handshaking signals. The *modem control register* is an 8-bit register that controls the output handshaking signals. The bit location and meaning of the output handshaking signals are identified in Figure 10.15.

In addition to the RTS and DTR previously described, there are three additional signals available. Two of these are the *out signals*. These lines are not used in a IBM serial adapter board. However, in other applications these two outputs could be used to control external functions such as radio or modem transmitter relays, channel switches, or similar devices. The third output from the register is called the *loop*. A logic 1 written to this bit places the 8250 in a *loop-back mode*. When operating in this mode the TxD line is connected to the RxD line. Also the RTS line is connected to the CTS line, the DTR is connected to the DSR, and out 1 and out 2 are connected to RLSD and RI inputs. The loop-back mode allows for testing of the hardware.

Just as the modem control register allowed the programmer to set the output handshaking lines, the *modem status register* (MSR) provides the programmer the ability to inspect the input handshaking lines. Figure 10.16 shows a bit map of the MSR register.

Bit 0. A logic 1 in this bit location means that there was a change in the *clear to send input* since this bit was last read.

Bit 1. Like bit 0, a logic 1 in this location means that there was a change in the *data set ready input* signal.

Bit 2. This bit lets the processor know that the *ring indicator* has changed from a logic 1 to a logic 0 condition.

Bit 3. A logic 1 in this location means that there was a change in the *receive line signal detect* since the last time that MSR was read.

Bit 4. This bit is the complement of the *CTS input* which originates with the external receiving device.

Bit 5. This bit is the complement of the *RTS input*.

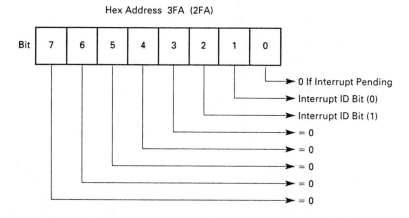

Figure 10.14 Interrupt Identification Register

Interrupt ID Register			INTERRUPT SET AND RESET FUNCTIONS			
Bit 2	Bit 1	Bit 0	Priority Level	Interrupt Type	Interrupt Source	Interrupt Reset Control
0	0	1	–	None	None	–
1	1	0	Highest	Receiver Line Status	Overrun Error or Parity Error or Framing Error or Break Interrupt	Reading the Line Status Register
1	0	0	Second	Received Data Available	Receiver Data Available	Reading the Receiver Buffer Register
0	1	0	Third	Transmitter Holding Register Empty	Transmitter Holding Register Empty	Reading the IR Register (if Source of Interrupt) or Writing into the Transmitter Holding Register
0	0	0	Fourth	Modem Status	Clear to Send or Data Set Ready or Ring Indicator or Received Line Signal Direct	Reading the Modem Status Register

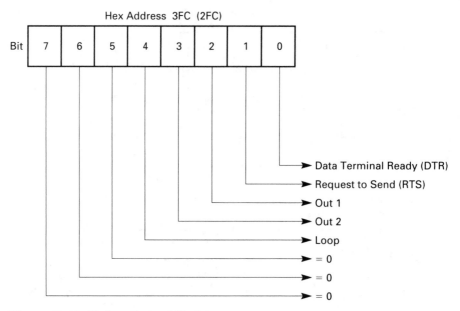

Figure 10.15 Modem Control Register

Bit 6. This is the complement of the *ring indicator input*.

Bit 7. This bit is the complement to the *receive line signal detect* input to the 8250 chip.

PROGRAMMING THE 8250

After identifying the functions of the various internal registers in the 8250 focus can be directed towards controlling the functions of the chip. The chip must be initialized as a serial port. This process consists of two steps. First, the programmer must set the divisor latches so that the baud rate can be established. Secondly, the number of data bits and stop bits must be selected along with the type of parity. The following code segments shown in C and BASIC will set the 8250 (serial port #1) for 1200 baud, eight data bits, one stop bit, and no parity.

After the port has been initialized, communications can begin between the sending device and the receiving device. Prior to delivering a character to the transmitter for sending the line status register (LSR) should be checked to make sure that the *transmit holding register* (THR) is empty. If a character is loaded in the THR and another character is moved into the register the first character will be overwritten. The following code examples perform a check of the THR. When the THR is empty then an output is performed.

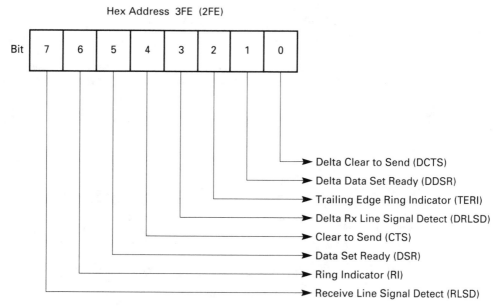

Figure 10.16 Modem Status Register

The line status register can also be used to verify if there is received data waiting to be read. The following code examples illustrate the checking of bit 0 of the LSR to determine if a character has been received but not retrieved. If there is a character waiting in the routine will get the character and print it on the screen.

The basics of controlling the port initialization, transmission and reception of data have been presented. Controlling the handshaking signals is dependent upon the logic of the program design and the software. The mechanics of activating a particular out are rather simple. To activate a particular bit, set the bit to a logic 1 at the appropriate modem control register. To check the status of a handshaking line the process involves masking and is performed similarly to the masking accomplished in earlier chapters.

INTERFACING PROBLEMS

Problem #1: Initialize the serial port to 300 baud, eight data bits, two stop bits, and no parity. After initializing the port use the serial port test block to monitor the DTR and RTS lines. Alternately set each of these two lines high and the other line low.

Problem #2: Initialize the serial port to 50 baud, seven data bits, two stop bits, and even parity. Transmit characters typed at the keyboard to the serial port. Verify that the TxD line "flickers" as the characters are being transmitted over the TxD line.

```
/* First set the BAUDRATE to 1200      */

baud_rate = 1843200/(1200*16);           /* baud.rate = divisor    */
outp(0x3FB,0x80);                        /* Set DLAB to 1          */
outp(0x3F8,(baud_rate & 0xff));          /* Load LSB into latch    */
outp(0x3F9,((baud_rate & 0xFF00)>>4));   /* Load MSB into latch    */

' Now set the format
outp(0x3FB,3);                            /* 8, 1, and NO Parity    */



' First set the BAUDRATE to 1200

baud.rate = 1843200/(1200*16)            ' baud.rate = divisor
OUT &h3FB,&h80                           ' Set DLAB to 1
OUT &h3F8,(baud.rate AND 255)            ' Load LSB into latch
OUT &h3F9,((baud.rate AND &hFF00)/256)   ' Load MSB into latch

' Now set the format
OUT &h3FB,3                               ' 8, 1, and NO Parity
```

Figure 10.17 Setting Serial Port using C and BASIC

```
/* Loop until THRE    */
while (!((status = inp(0x3fd)) & 32));  /* while bit 5 is low ... */

outp(0x3F8,var);
```

```
' Loop until THRE
DO
  STATUS = INP (&H3FD)
LOOP UNTIL ((STATUS AND 32) = 32)

' Now the Transmitter is ready
OUT &H3F8,VAR
```

Figure 10.18 C and BASIC Code Examples Checking THR

```
/* Check for flag    */
if ((status = inp(0x3fd)) & 1))          /* if bit 0 is hi ...          */
  var = inp(0x3f8);                      /*    get the character        */
```

```
' Check for flag
STATUS = INP(&h3fd)
IF ((STATUS AND 1) = 1) THEN  var = inp(0x3f8);       '    get the character
```

Figure 10.19 C and BASIC Code Examples Checking LSR

Problem #3: Construct the proper interface cable so that two computers are connected through a serial cable. Do not use handshaking! Write the necessary software so that characters typed on a computer are displayed on the other computer and vice versa.

Problem #4: Modify the cable and software used in Problem #3 so that before the computer sends a character the receiving computers RTS line must be at a logic 1 level. Write the software so that as long as this line is held a logic 1 characters typed on the keyboard will be transmitted to the other computer and displayed. As a character is transmitted the character should be echoed so that it is displayed on the transmitting computer screen as well as the receiving computer screen.

CHAPTER 11

Timers and Counters

Time can play an important role when a computer is used for data acquisition and control. Chapter 4 was devoted to studying the computer's time-of-day clock and to the use of time as a variable in interfacing and data acquisition projects and applications. When thinking of time in the way that it was presented in Chapter 4, time is viewed in absolute terms; some task is performed under the control of the computer at a specific time of day. Not all time measurements are in absolute terms. For example, if a cake recipe calls for a baking time of 30 minutes, it doesn't matter what time of day that the cake is started, but what is important is the number of minutes that the cake is in the oven. In this illustration, the measure of time is relative and is not necessarily dependent on a time-of-day measurement.

A measure of *relative time* is the difference between the *start time* and the *end time* of a process or procedure. In the cake baking example, the baking of the cake can follow either of the following two procedures.

1. (a) Determine the time of day and add 30 minutes; calculating the "take-out" or ending baking time.

   (b) Watch the clock.

   (c) When the time of day equals the "take-out" time, remove the cake from the oven.

or

2. (a) Set a stove timer for 30 minutes.

   (b) Do something fun until the timer goes off.

   (c) When the timer sounds, remove the cake from the oven.

The first procedure is the technique that was described in Chapter 4, where the system time-of-day clock played an important role in controlling electrical circuits, machinery, or other devices according to time of day. If this method of timekeeping is used in the cake baking example, the computer must continually access the computers system clock, compare the current time to the intended baking stop time, and determine if the stop time has been reached or exceeded. To be certain that the stop time wasn't missed, the program should be written to keep the software loop repeatedly checking the time of day as frequently as possible.

This programming strategy, while effective, has an important disadvantage. While the program is constantly "watching" the clock, the program and, therefore, the computer can't do anything else. The computer is fully committed to doing one thing, watching the clock and only watching the clock. This method if programming, which in essences keeps the computer focused on the clock, is an inefficient use of computer resources. This strategy also places a high level of importance on using a computer with a high-speed clock along with writing the program as efficiently as possible.

The time-keeping method described in the second scenario is a better choice for a couple of reasons. In the second scenario, another hardware component (a *timer board*) in the computer system is given the responsibility of "watching" the baking time and letting the computer system know when the specified baking time has expired. By transferring the timing responsibility to another hardware system in the computer, the microprocessor is free to go on with other tasks while the timer board is responsible for the baking process. The freedom to start a timed event, turn to another task, and later return to the timed process is a great advantage in busy real-world applications.

Chapter 4 had an example in which a burglar alarm was controlled by a time-of-day clock. The clock was used to arm the alarm system during the hours that the business was closed, presumable the evening hours. In this example, if the design of the software program required the alarm bell to be sounded for 10 minutes after a burglar entry was detected, the programmer would have two alternate ways of controlling the bell.

One method is to use the time-of-day clock. The programmer could write the program so that at the instant the burglar was detected, 10 minutes would be added to the time-of-day clock. The computer would then start the alarm bell ringing and begin checking the system time-of-day clock for the correct turn OFF time which is 10 minutes after the alarm was turned ON. If this technique was chosen, the computer would be limited in what other functions could be performed while it was waiting for the 10 minutes to pass.

Another time-keeping technique involves the use of an elapsed timer. In this technique, a timer board in the computer directly controls electrical power to the burglar alarm bell. In this approach to the burglar alarm example, when the computer detected a burglar in the building the alarm bell would be turned ON, and the 10-minute timer would be started. The bell would be under control of the timer and when 10 minutes had expired the timer would turn OFF the bell. This technique, using a separate timer, allows the computer to be free to monitor security in other parts of the building, activate a modem to call the police, turn ON lights, and in other ways control the security of the building. Obviously, for this application a separate timer is a better choice than relying on the computer systems time-of-day clock.

Whether time is measured in absolute (time-of-day) terms, or relative (elapsed time), depends on the situation and the intent of the program. The decision to use either form of time measurement is often the responsibility of the programmer, engineer, or technician. Data acquisition programming often requires both techniques for measuring time. Absolute time is best measured through the system time-of-day clock, and relative time is best measured through the use of one or more timers.

One way of visualizing timers is to think of them as digital counters whose purpose is to count seconds of elapsed time. In other words, if the input to a counter is a pulse every tenth of a second, the counter functions as a timer with a resolution of one-tenth of a second. In this example the source of the pulse is presumed to be an oscillator that produces 10 pulses per second.

Timers can also function as an event counter. When timers are used as an event counter, the intent of the counter is to count input pulses from physical devices such as a limit switch, photo-detector, or other physical device. The source of the incoming pulse is really the determining factor whether a timer is working as an elapsed time monitoring device (timer) or whether the timer circuit is serving as a counter. If the source of the pulses is a repetitive oscillator, the timer circuit is serving as a tuner. If the source of the pulse is an external device or other nonperiodic source of pulses, the timer circuit is serving as a counter. Sometimes, application of a timer requires careful analysis to determine the true functional role that the timer circuit is serving.

This flexibility is found with expansion *timer/counter boards*. They, like other boards, plug into expansion slots in the PC bus. The CTM-05 produced by Keithley MetraByte, the CIO-CTR05 manufactured by Computer Boards, Inc., or the AD1000 produced by Real Time Devices, Inc., P.O. Box 906, State College, PA 16804, are examples of timer boards that can be installed in the IBM-PC bus. Each board differs in specific ways, but they do share some characteristics. One characteristic that these boards share is the ability to measure elapsed time because of an on-board oscillator. The boards also share the flexibility of being configured as an event counter if input pulses are applied to the board from an external source.

The Keithley MetraByte CTM-05, officially known as a Multi-Function Counter Timer Digital Expansion Board, is illustrated and described in Figure 11.1. The CTM-05 board is based on the Advanced Micro Devices AMD-9513 System Timing Controller integrated circuit. This chip is the "heart" of the CTM-05 as it provides the timing function for the expansion board. The board has five 16-bit up/down counters, a 1 MHz crystal oscillator, and a programmable divider circuit. Figure 11.2 shows a block diagram of the major elements of the CTM-05 Timer Board.

Installation of the CTM-05 is similar to the installation of any other interfacing boards. Physically, the board can reside in any of the expansion slots in the system or mother board. However, before the board is actually installed, the base address must be set and the level of hardware interrupt must be chosen.

As was discussed with other interfacing boards each hardware device must have a unique port address. The I/O port map shown in Figure 2.3 should be consulted. Like the other boards the addresses in the range, 300 Hex through 31F

**5-CHANNEL
COUNTER-TIMER BOARD**

CTM-05

FEATURES

- 5 independent 16-bit counters
- Uses industry standard 9513 chip
- Counts frequency to 7 MHz
- Up/down and binary/BCD counting
- Internal 1-MHz frequency source
- Programmed frequency output
- Complex duty cycle outputs
- Alarm comparators on counters 1 and 2
- One-shot or continuous outputs
- Retriggering capability
- Programmable count/gate source selection
- Programmable input/output polarities
- Programmable gating functions
- 8-bit latched input port
- 8-bit latched output port
- Interrupt input channel
- Software included

APPLICATIONS

- Event counting
- Frequency synthesis
- Coincidence alarms
- Frequency shift keying (FSK)
- Complex pulse generation
- Retriggerable digital timing functions
- Waveform analysis

FUNCTIONAL DESCRIPTION

Keithley MetraByte's CTM-05 5-Channel Counter-Timer board, plugs into the IBM PC/XT/AT and compatibles. It provides five general purpose 16-bit counters. A selection of various internal frequency sources and outputs can be selected as inputs for individual counters. Counter inputs are software-selectable active-high or active-low. Each counter may be gated in hardware or by software. The counters can be programmed to count up or count down in either binary or BCD. All five counters can be connected together by software to form a 32, 48, 64 or 80 bit counter.

Each counter has a Load Register and a Hold Register. The Load Register is used to automatically reload the counter to a programmed value, thus controlling the count and count period. The Hold Register is used to save count values without disturbing the counting process. This permits the computer to read intermediate counts. The Hold Register may also be used as a second Load Register to generate complex waveforms.

Figure 11.1 Photograph and Specifications for CTM-05 Timer Board (Courtesy Keithley MetraByte Corporation)

BLOCK DIAGRAM

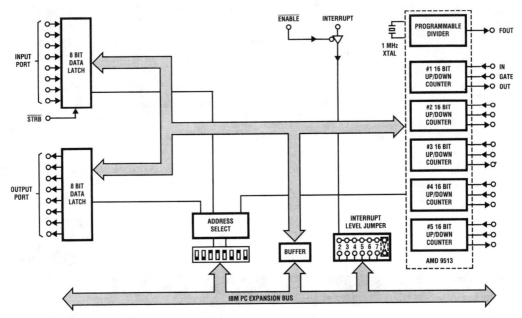

Figure 11.2 CTM-05 Block Diagram (Courtesy Keithley MetraByte Corporation)

Hex are excellent choices for addressing interfacing boards and related expansion boards like the CTM-05 Timer.

For the purpose of this example and the programming examples later in this chapter, 300 Hex has been chosen as the base address of the timer board. Naturally, the selection the board address must be done with care to avoid address locations that have already been assigned to other boards. Address configuration is accomplished by the setting DIP switches. These switches correspond to the address lines A2 through A9 and are marked so that A9 is the left most switch. To select 300 Hex as the base address, place the A9 and A8 switches in the down position and all others in the up position. Figure 11.3 shows the correct switch settings for a base address of 300 Hex.

The next step in the board installation process is to select the level of hardware interrupt that the timer board will generate when it wants attention from the microprocessor. Use of the interrupt capability of the microprocessor allows the timer to do a timing function while the microprocessor is doing something else. When the timer completes a task, the processor is interrupted, and in essence, the processor is notified that the timer is finished. Selection of the interrupt level

BASE ADDRESS

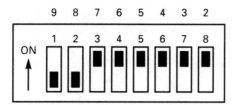

(a) Base Address Switch Set to 300 Hex

Switch	Address line	Value when Switch is OFF	
		Decimal	Hex
1	A9	512	200
2	A8	256	100
3	A7	128	80
4	A6	64	40
5	A5	32	20
6	A4	16	10
7	A3	8	8
8	A2	4	4

(b) Base Address Switch Values

Figure 11.3 Setting Base Address of 300 Hex

is important as more than one device generating the same interrupt should be avoided.

If there is doubt about selecting a level of interrupt, the "X" option should be chosen. This interrupt option on the jumper block disables the interrupt function.

PROGRAMMING THE CTM-05

There are many uses for a counter/timer board. One example of a clever use of a timer has been described in the hypothetical burglar alarm application. We have already mentioned that this board could control the alarm bell. In this use, the computer would essentially cause the CTM-05 board to turn ON the bell while the computer attended to something else. This example uses the counter/timer as a programmable monostable multivibrator, also known as a one-shot device. In this example, the programmer would need to configure the board as a timer that made its output voltage high (+ 5 volts) from the beginning of the timing cycle until the time had expired. At the conclusion of the timing cycle the output voltage would return to the original starting value of zero volts. As the output from the timer board goes through the cycle, beginning at a low logic level, outputting a high logic level for a period, and returning to a low logic level, it clearly exhibits the characteristics of a one-shot multivibrator.

Another use for the CTM-05 board is as a pulse generator. For example, an external circuit needs a repetitive pulse every ten milliseconds. This type of pulse can be provided by the computer through a counter/timer board. Once configured and started, the board would output a repetitive waveform in which the duty cycle would produce a positive pulse every 10 milliseconds.

The CTM-05 can also be configured to provide a bistable output. In this mode of operation the board could produce waveform with a 50% duty cycle. For example, the waveform may have a one-second ON period followed by a one-second OFF period. This output from the CTM-05 can be continuous, leaving the computer free to do another task.

These are just a few of the possible uses of the CTM-05 Timer Board. In the examples provided, the board was used as a timer. However, if programmed properly, the board can serve as a counter, monostable multivibrator, or bistable multivibrator. Programming the CTM-05 can be done at the chip level or through the use of the software that is shipped by Keithley MetraByte Inc. with the board. Since this library is available at no extra cost, the programming examples will concentrate on this method of setting up the CTM-05. The software library on the disk, CTM5.LIB, includes the functions shown in Figure 11.4. These functions can be linked with programs as they are developed. Support is available in this library for the interpreted language BASIC, also Borland C, Microsoft C, Turbo Pascal, and Microsoft Fortran compilers.

Figure 11.4 shows the programming "modes" for the CTM-05. These modes are simply the types of commands that can be sent to the board. For example, mode 0 is used to initialize the board while mode 1 is used to set the counter mode control registers. In the following paragraphs, these commands will be explained in some detail. However, because of the complexity of the CTM-05, it may be necessary to refer to the Keithley MetraByte CTM-05 User's Manual for specific information.

The programmer controls the CTM-05 through calls to the software driver. These calls tell the driver about several important characteristics. This information

Mode	Function
0	Initialize
1	Set a counter Mode Register
2	Multiple Counter Control commands
3	Load selected Counter Load Register
4	Read a selected Counter Hold Regoster
5	Read Digital Input Port
6	Write Digital Output Port
7	Latch counter(s) and store data on interrupt
8	Return status of interrupts
9	Unload interrupt data from memory
10	Measure frequency
11	Latch counters and store segment and offset

Figure 11.4 CTM-05 Timer Board Modes

is sent to the driver in a standardized format. This format, or calling convention, for Turbo C (large memory model) is:

```
int mode, parms[2.], flags;
...
TCL_CTM5(&mode, parms, &flags);
```

If the reader is programming in Turbo C, he/she must use a project file so that the CTM-05 library can be linked to the program. Once the project includes the source file, and the library, Turbo C knows where to find everything that is needed. If the reader is using Quick BASIC, it is important to remember to invoke the program with the command:

```
qb /ctm5.qlb
```

while the CTM5.QLB library in the default directory. After this is done, to use driver functions such as initialize, the programmer must specify the *function* (the mode), load any necessary *parameters* (parms), and provide a place for *return code* (flags). When the function is completed, the return value in *flags* will tell the program if the operation was successful. As each of the driver functions is discussed, possible entry conditions and return codes will be presented.

Mode 0: Initialize Driver

This mode initializes the software driver by telling it the port address of the board, the source of the signal on output FOUT, and the FOUT divider ratio. *FOUT* is the programmable divider that is connected to the 1 MHz crystal.

Mode = 0
Parms[0] = Base address (100 Hex to 3FC Hex)
Parms[1] = FOUT divider ratio (0 to 15) 0 = /16 else N = /N
Parms[2] = FOUT source
 0 = F1 (Master clock = 1 MHz)
 1 = Source 1
 2 = Source 2
 3 = Source 3
 4 = Source 4
 5 = Source 5
 6 = Gate 1
 7 = Gate 2
 8 = Gate 3
 9 = Gate 4
 10 = Gate 5
 11 = F1
 12 = F2 (F1 / 10 = 100 KHz)
 13 = F3 (F1 / 100 = 10 KHz)
 14 = F4 (F1 / 1000 = 1 KHz)
 15 = F5 (F1 / 10000 = 100 Hz)
Parms[3] = Compare 2 enabled/disabled 0/1
Parms[4] = Compare 1 enabled/disabled 0/1
Parms[5] = Time-of-day mode (0 to 3)

Return Values

Flags = 0	No error
2	Mode number out of range <0 or >11
3	Base address out of range
11 through 19	Parms[n] out of range

Mode 1: Set Counter Mode Register

Mode 1 sets a counter by setting its mode register. Following setting mode 0, this is the next logical command.

Entry Requirements

Mode = 1
Parms[0] = Counter number (1 to 5)
Parms[1] = Gating control (0 to 7)
     0 = No gating
     1 = Active high level TCN-1

Mode 1: Entry Requirements Continued

---

       2 = Active high level gate $N+1$
       3 = Active high level gate $N-1$
       4 = Active high level gate N
       5 = Active low level gate N
       6 = Active high edge gate N
       7 = Active low edge gate N
Parms[2] = Count edge positive/negative 0/1
Parms[3] = Count source selection (0 to 15)
       0 = TCN
       1 = Source 1
       2 = Source 2
       3 = Source 3
       4 = Source 4
       5 = Source 5
       6 = Gate 1
       7 = Gate 2
       8 = Gate 3
       9 = Gate 4
     10 = Gate 5
     11 = F1
     12 = F2
     13 = F3
     14 = F4
     15 = F5
Parms[4] = Disable/enable special gate 0/1
Parms[5] = Reload from LOAD/LOAD OR HOLD 0/1
Parms[6] = Count once/count repeatedly 0/1
Parms[7] = Binary count/BCD count 0/1
Parms[8] = Count down/count up 0/1
Parms[9] = Output control (0 to 5 except 3)
       0 = Inactive, output low
       1 = Active high terminal count pulse
       2 = Terminal count toggled
       3 = Illegal
       4 = Inactive, output high impedance
       5 = Active low terminal count pulse

Return Values

Flags =	0	No error
	1	Base address unknown
	2	Mode number out of range <0 or >11
	11 through 19	Parms[n] out of range

Mode 2: Multiple Counter Control Commands

This mode enables the program to load, latch and save, enable (arm), and disable (disarm), individual or multiple counters.

Entry Requirements

Mode = 2	
Parms[0] =	Operation (1 to 6)
	1 = Arm selected counter(s)
	2 = Load counter to source
	3 = Load and arm counter
	4 = Disarm and save counter
	5 = Latch counter to hold register
	6 = Disarm counter
Parms[1] =	Select counter 1 0/1
Parms[2] =	Select counter 2 0/1
Parms[3] =	Select counter 3 0/1
Parms[4] =	Select counter 4 0/1
Parms[5] =	Select counter 5 0/1

Return Values

Flags = 0	No error
1	Base address unknown
2	Mode number out of range <0 or >11
10	Command number out of range <1 or >6
11 through 15	Counter n not 0 or 1

Mode 3: Load Counter Load Register

This mode is used to place a value in a counter's load register. The value is not loaded into the counter (see mode 2).

Entry Requirements

Mode = 3	
Parms[0] =	Counter number (1 to 5)
Parms[1] =	Load data (-32768 to $+32767$)

Return Values

Flags = 0	No error
1	Base address unknown
2	Mode number out of range <0 or >11
10	Counter number out of range <1 or >5

Mode 4: Read Selected Counter Hold Register

This mode is used if it is necessary for a software program to know the count in a specific counter. If this information is necessary, mode 2 must be used to *latch counter to hold register*. Once this is done mode 4 can be used to read the value in the hold register. There is no way to directly read the contents of a counter.

Entry Requirements

Mode = 4
Parms[0] = Counter number (1 to 5)
Parms[1] = Data read variable, value does not matter

Return Values

Parms[0] = Counter number (0 to 5)
Parms[1] = Counter data (-32768 to $+32767$)
Flags = 0   No error
        1   Base address unknown
        2   Mode number out of range <0 or >11
        10   Counter number out of range <1 or >5

Mode 5: Read Digital Input Port

The CTM-05 has an 8-bit digital port built into the board. Mode 5 allows the programmer to read a digital value at the input port. The incoming data will be eight binary bits.

Entry Requirements

Mode = 5
Parms[0] = Data read variable, value does not matter

Return Values

Parms[0] = Input port data (0 to 255)
Flags = 0   No error
        1   Base address unknown
        2   Mode number out of range <0 or >11
        10   Command number out of range <1 or >6
        11 through 15 Counter n not 0 or 1

Mode 6: Write Digital Output Port

Similar to mode 5, this mode enables the programmer to use the digital I/O port that is part of the CTM-05 board. Mode 6 writes an 8-bit digital value to the output port.

Entry Requirements

Mode = 6	
Parms[0] =	Value to be output

Return Values

Flags =	0	No error
	1	Base address unknown
	2	Mode number out of range <0 or >11
	10	Output data out of range <0 or >255

Mode 7: Latch Counters and Save on Interrupt

Mode 7 (also modes 8 and 11) use the computers interrupt system to perform their tasks. Interrupts are a powerful tool, however, they are beyond the scope of this book. Therefore, only a brief description of these modes will be given. The authors suggest that interrupts be investigated as the reader becomes more experienced in programming. Mode 7 transfers the contents of selected counters to buffer memory every time an interrupt occurs.

Entry Requirements

Mode = 7	
Parms[0] =	Number of interrupts (1 to 32767)
Parms[1] =	Memory segment for counter 1 (0 to 65535)
Parms[2] =	Memory segment for counter 2 (0 to 65535)
Parms[3] =	Memory segment for counter 3 (0 to 65535)
Parms[4] =	Memory segment for counter 4 (0 to 65535)
Parms[5] =	Memory segment for counter 5 (0 to 65535)
Parms[6] =	Start on IP0 disabled/enabled 0/1
Parms[7] =	Interrupt level (2 to 7)

Return Value

Flags =	0	No error
	1	Base address unknown
	2	Mode number out of range <0 or >11
	10	Interrupt level out of range <2 or >7

Mode 8: Return Status of Interrupts

This mode returns a value that describes the process started in mode 7.

Entry Requirements

Mode = 8	

Return Values

Parms[1] =	Interrupt active/finished 1/0

Mode 8: Return Values Continued

Parms[2] = Current word count
Flags = 0   No error
      1   Base address unknown
      2   Mode number out of range <0 or >11

Mode 9: Transfer Data During/After Interrupt

Mode 9 is a general purpose block transfer routine. With this mode data can be copied from one place in memory to a properly dimensioned array.

Entry Requirements

Mode = 9
Parms[0] = Number to words to transfer (1 to 32767)
Parms[1] = Starting word number (0-parms[0])
Parms[2] = Memory segment to transfer from
Parms[3] = Offset of target array

Return Values

Flags = 0   No error
      1   Base address unknown
      2   Mode number out of range <0 or >11
    10   Number of words <=0
    11   Starting word number <0

Mode 10: Measure Frequency

This mode uses several features of the AMD9513 timer chip to measure up to nine external frequency inputs. These inputs (TTL compatible) should be connected to any of the source inputs or gates 1 through 4.

Entry Requirements

Mode = 10
Parms[0] = Gate interval (1 to 32767)
Parms[1] = Select input signal Source (1 to 9)
      1 = Source 1
      2 = Source 2
      3 = Source 3
      4 = Source 4
      5 = Source 5
      6 = Gate 1
      7 = Gate 2
      8 = Gate 3
      9 = Gate 4

Return Values

Parms[2] = Count accumulated during gating interval

Mode 10: Return Values Continued

Flags	=	0	No error
		1	Base address unknown
		2	Mode number out of range <0 or >11
		10	Gating interval out of range <1 or >32767
		11	Source input out of range <1 or >9

Mode 11: Latch Counters and Save on Interrupt—Data to Memory

Mode 11 is similar to mode 7 except that the programmer can pass the offset address and segment address of the data byte. This mode can only be used after mode 7 has been called.

Entry Requirements

Mode = 11
Parms[0] = Memory offset for a counter 1 data dump
Parms[1] = Memory segment for a counter 1 data dump
Parms[2] = Memory offset for a counter 2 data dump
Parms[3] = Memory segment for a counter 2 data dump
Parms[4] = Memory offset for a counter 3 data dump
Parms[5] = Memory segment for a counter 3 data dump
Parms[6] = Memory offset for a counter 4 data dump
Parms[7] = Memory segment for a counter 4 data dump
Parms[8] = Memory offset for a counter 5 data dump
Parms[9] = Memory segment for a counter 5 data dump

Return Values

Flags	=	0	No error
		1	Base address unknown
		2	Mode number out of range <0 or >11
		10	Interrupt count out of range
		17	Interrupt level not between 2 and 7
		21	Counter 1 segment wraparound
		22	Counter 2 segment wraparound
		23	Counter 3 segment wraparound
		24	Counter 4 segment wraparound
		25	Counter 5 segment wraparound

USING THE DRIVER FUNCTIONS

Following a description of the individual software driver functions, a few program examples will be provided. The program example shown in Figure 11.5 illustrates six calls to the driver. The first call is mode = 0, and it initializes the driver, telling it the base address of the board (parms[0]), the source of the count (parms[2]),

```
#include <stdio.h>
#include <conio.h>
#include <process.h>

extern TCL_CTM5(int*, int*, int*);        /* proto type for driver calls */

void main(void)
   {
     int mode, parms[15], flags;
     unsigned i;

     /*  Initialize the driver    */
     mode = 0;
     parms[0] = 0x300;          /* Base address                    */
     parms[1] = 1;              /* Divide by one                   */
     parms[2] = 12;             /* F2 is the source                */
     parms[3] = 0;              /* Disable Compare 2               */
     parms[4] = 0;              /* Disable Compare 1               */
     parms[5] = 0;              /* Disable Time of day mode        */

     TCL_CTM5(&mode,parms,&flags);  /* Make the call               */
     if(flags !=0)                  /* Check for success           */
        {
          printf("ERROR Initializing driver  %d\n\n",flags);
          exit(1);
        }

     /*  Initialize Counter 1    */
     mode = 1;
     parms[0] = 1;              /* Just counter #1                 */
     parms[1] = 0;              /* No gating                       */
     parms[2] = 0;              /* Positive edge                   */
     parms[3] = 12;             /* F2 is source                    */
     parms[4] = 0;              /* Disable Special Gate            */
     parms[5] = 0;              /* Reload from load register       */
     parms[6] = 0;              /* Count once only                 */
     parms[7] = 0;              /* Use Binary counting             */
     parms[8] = 1;              /* Count up                        */
     parms[9] = 0;              /* Output is inactive              */


     TCL_CTM5(&mode,parms,&flags);  /* Make the call               */
     if(flags !=0)                  /* Check for success           */
        {
          printf("ERROR Initializing Counter Mode register  %d\n\n",flags);
          exit(1);
        }

     /*  Load the Load register with 0    */
     mode = 3;
     parms[0] = 1;              /* Counter number 1                */
     parms[1] = 0;              /* Preload with 0                  */
```

(a)

Figure 11.5 Programming Example Showing Driver Calls

```
TCL_CTM5(&mode,parms,&flags);  /* Make the call                  */
if(flags !=0)                  /* Check for success              */
  {
    printf("ERROR setting Load register  %d\n\n",flags);
    exit(1);
  }


/*  Load the counter from the Load Register   */
mode = 2;
parms[0] = 3;                  /* Loadand arm counter            */
parms[1] = 1;                  /* Select Counter 1               */
parms[2] = 0;                  /* Disabled                       */
parms[3] = 0;                  /* Disabled                       */
parms[4] = 0;                  /* Disabled                       */
parms[5] = 0;                  /* Disabled                       */

TCL_CTM5(&mode,parms,&flags);  /* Make the call                  */
if(flags !=0)                  /* Check for success              */
  {
    printf("ERROR Loading data  %d\n\n",flags);
    exit(1);
  }

for (i=0;  i<65000;  i++);  /* Delay                             */


/*  Latch the counter into the Hold Register   */
mode = 2;
parms[0] = 5;                  /* Latch counter                  */
parms[1] = 1;                  /* Select Counter 1               */
parms[2] = 0;                  /* Disabled                       */
parms[3] = 0;                  /* Disabled                       */
parms[4] = 0;                  /* Disabled                       */
parms[5] = 0;                  /* Disabled                       */

TCL_CTM5(&mode,parms,&flags);  /* Make the call                  */
if(flags !=0)                  /* Check for success              */
  {
    printf("ERROR Latching data  %d\n\n",flags);
    exit(1);
  }

/*  Read the Hold Register   */
mode = 4;
parms[0] = 1;                  /* Select counter                 */
parms[1] = 1;                  /* Dummy value                    */

TCL_CTM5(&mode,parms,&flags);  /* Make the call                  */
if(flags !=0)                  /* Check for success              */
  {
    printf("ERROR Latching data  %d\n\n",flags);
    exit(1);
  }
else
  {
    printf("\n\nThe data read was %u\n\n",parms[1]);
  }
}
```

(b)

Figure 11.5 Programming Example Showing Driver Calls

how to further divide the source (parms[1]), and disabling compare 2, 1, and time-of-day mode. If there are no errors the program proceeds to the second call.

The second driver function call is mode = 1. Recall that mode 1 is used to set a specific counter's mode register. In this example, the comments after each line of code greatly simplify the assignments. Mode 3 tells the board to start at 0 when it counts. Once the driver and the counter have been initialized, the call to mode 2 starts the counter running. In this example, only counter 1 is *loaded* and *armed*. After a little delay (the *for* loop), a call to mode 3 latches the contents of the counter to the holding register so that the value in the counter can be read when mode 4 (*read selected holding register*) is called.

The next program example (see Figure 11.6) assumes that a computer will control a pneumatic ram. This ram is supposed to be extended for ten seconds and then retracted for ten seconds. This procedure of extension and retraction should continue continuously. Notice that just as in the previous example, mode 0 is the first mode to be called. In this example though, the mode 1 call is different. Parms[6] is now set to count repeatedly, parms[8] is set to count down, and parms[9] is set to toggle the output every time the counter gets to 65535. Finally, the value in the load register is set to 1000. This is the *load value*. At a speed of 100 Hz, 1000 counts would equal 10 seconds. If the output from counter 1 was used to control a two-position electric valve, the RAM would work as the specification called for.

PROGRAMMING PROBLEMS

Problem #1. Write a program that causes the counter/timer board to generate a positive going pulse every 10 milliseconds.

Problem #2. Write a program and wire the necessary external circuitry so that the counter/timer board counts passing objects and displays the count on the screen. For example, an infrared transmitter/receiver could count marbles as they roll on a track.

Note: If the IR devices are adjusted to the proper height, they will be able to count the marbles even if the marbles are touching when they pass the IR detector.

Problem #3. In the previous assignment, it is clear that when the marbles are moving quickly it is difficult to count and, therefore, control them. If it is important to count the marbles so that an accurate number can be loaded into boxes, it might be helpful to space the marbles out before they pass the IR detector. The flow of objects down a line can be controlled with capstans just as the capstans (motor driven rollers) control audio and video-tape. A soft foam capstan on either side of the track can control the flow of marbles. Write a program and wire the necessary circuitry to space the marbles one second apart.

```c
#include <stdio.h>
#include <conio.h>
#include <process.h>

extern TCL_CTM5(int*, int*, int*);        /* proto type for driver calls */

void main(void)
  {
    int mode, parms[15], flags;
    unsigned i;

    /*  Initialize the driver   */
    mode = 0;
    parms[0] = 0x300;              /* Base address                       */
    parms[1] = 1;                  /* Divide by one                      */
    parms[2] = 12;                 /* F2 is the source                   */
    parms[3] = 0;                  /* Disable Compare 2                  */
    parms[4} = 0;                  /* Disable Compare 1                  */
    parms[5] = 0;                  /* Disable Time of day mode           */

    TCL_CTM5(&mode,parms,&flags);  /* Make the call                      */
    if(flags !=0)                  /* Check for success                  */
      {
        printf("ERROR Initializing driver  %d\n\n",flags);
        exit(1);
      }

    /*  Initialize Counter 1   */
    mode = 1;
    parms[0] = 1;                  /* Just counter #1                    */
    parms[1] = 0;                  /* No gating                          */
    parms[2] = 0;                  /* Positive edge                      */
    parms[3] = 15;                 /* F5 is the source                   */
    parms[4] = 0;                  /* Disable Special Gate               */
    parms[5] = 0;                  /* Reload from load register          */
    parms[6] = 1;                  /* Count repeatedly                   */
    parms[7] = 0;                  /* Use Binary counting                */
    parms[8] = 0;                  /* Count down                         */
    parms[9] = 2;                  /* Output toggles                     */


    TCL_CTM5(&mode,parms,&flags);  /* Make the call                      */
    if(flags !=0)                  /* Check for success                  */
      {
        printf("ERROR Initializing Counter Mode register  %d\n\n",flags);
        exit(1);
      }

    /*  Load the Load register with 1000   */
```

(a)

Figure 11.6 Program Example for Pneumatic Ram

```
mode = 3;
parms[0] = 1;                    /* Counter number 1            */
parms[1] = 1000;                 /* Preload with 1000           */

TCL_CTM5(&mode,parms,&flags);   /* Make the call               */
if(flags !=0)                    /* Check for success           */
   {
     printf("ERROR setting Load register  %d\n\n",flags);
     exit(1);
   }


/*  Load the counter from the Load Register   */
mode = 2;
parms[0] = 3;                    /* Loadand arm counter         */
parms[1] = 1;                    /* Select Counter 1            */
parms[2] = 0;                    /* Disabled                    */
parms[3] = 0;                    /* Disabled                    */
parms[4] = 0;                    /* Disabled                    */
parms[5] = 0;                    /* Disabled                    */

TCL_CTM5(&mode,parms,&flags);   /* Make the call               */
if(flags !=0)                    /* Check for success           */
   {
     printf("ERROR Loading data  %d\n\n",flags);
     exit(1);
   }


/*   From this point on the timer runs without any more attention   */


}
```

(b)

Figure 11.6 Program Example for Pneumatic Ram

In an actual industrial control application, this pulse would be connected to the capstan or other electromechanical device so that the marbles would be evenly spaced as they roll toward the counting and boxing mechanism.

Problem #4. In Problem #3, the task was to evenly space the marbles under the control of a timer. Now, it is possible to count the marbles and to direct them to different destinations for packaging. Write a program and wire the necessary external circuitry to count five marbles and then switch a gate so that the next five go to a different "box." After every five go by, the gate should toggle.

CHAPTER **12**

Trends

Identifying trends can be a risky undertaking in any field of study, and the fields of data acquisition, interfacing, and personal computers are no exception. However, a couple of observable innovations by equipment and software manufacturers clearly define trends and new products that the industry is heavily marketing. These two innovations—multifunction boards and industry standard real-time data acquisition software—lead in directions that other vendors and users of interfacing products are following. This chapter will take a closer look at both of these trends.

MULTIFUNCTION BOARDS

This text has introduced the reader to several interfacing boards that are designed to be installed in the IBM-PC bus. Each board presented in the text has served a unique and sole purpose. The PIO-12 digital I/O board was presented early in the text as a low-cost method of interfacing digital signals into or out of a personal computer. In Chapter 8, the DAS-4 analog-to-digital board was presented as an inexpensive way of converting analog signals to digital format. The 8-bit DAS-4 offered an entry level approach to analog to digital conversion, while the DAS-8 was presented as a 12-bit A/D converter offering increased resolution. These single-purpose boards were selected because of low cost, dedication to a particular function, and simplicity. Each of the boards selected for highlighting in the text were simple to understand and equally simple to program. The single-function characteristic of these boards is an attractive feature, but their single function also has limitations. A limiting factor is that when more than one single function I/O board is used in a computer, the number of slots needed in the computer's mother board may exceed the number of open or unused slots. For example, a hypothetical interfacing and control application might require the following input and output functions:

Required I/O Functions

---

8 Channels of Analog Input
1 Channel of Analog Output
8 Bits of Digital Output

An example of an application that would require this mix of analog and digital I/O boards is the sensing and control interface for a computer-controlled hot water heating system. Typically, the system would consist of a computer controlled boiler. The computer, interfaced to thermostats, would control the flow of hot water to radiators throughout the building that is being heated. The eight analog channels specified in this system are used to measure the analog temperature in eight different rooms being heated. A single analog output channel is used to control the flow of hot water from the boiler to the heating system through a proportional valve or variable speed pump motor.

The eight bits of digital output are used to control the on/off status of eight different fans, one fan in each room. The binary output by the computer will depend upon which room or rooms are "calling for" heat. If a room is calling for heat the proportional valve would be opened, circulating the hot water, and the fan in the cold room would turn on.

If this system were implemented using the single function I/O boards that have been described in the text, the following boards would likely be specified:

Required I/O Boards

---

1 DAS-4   8-Bit A/D Converter
1 DAC-02 D/A Converter
1 PIO-12   8-Bit Digital I/O Board

This selection of I/O boards would accomplish the described control application. However, the inclusion of these boards would require the use of three slots on the computers mother board (see Figure 2.5(c)). The need for three slots could create a problem in that there may not be three open slots available in the computer system. Considering the number of slots required for essential items such as a video monitor board, hard disk controller, floppy disk controller, modem, sound board, fax board, and an SCSI board, which is often used with a CD-ROM device, open slots in a computer are a premium.

Data Translation DT2811

The need for three slots to accommodate our hypothetical control problem can be reduced by using a multifunction board. As implied by the name, this type of I/O board incorporates more than one interfacing function on a single board. This means that the board may have analog-to-digital and digital I/O, as well as other functions on a single board. Combining more than one function on a single board reduces the need for slots in the computer's mother board. However, use of this type of board may increase the complexity of the board.

An example of a popular multifunction board, available at a reasonable cost, is the DT2811 produced by Data Translation, Inc. The block diagram for this board is shown in Figure 12.1. A review of this diagram reveals that the board offers eight multiplexed analog input channels, two independent analog output channels, and 16 digital TTL compatible I/O lines.

With these different functions on a single board, it is obvious that this single board can accomplish the functions of the three individual boards that were specified in the boiler control problem described above. Indeed, other functions are available on the DT2811 that are not required by the hypothetical application. These unused functions, eight digital I/O bits and one analog output channel, could be used for future expansion of the proposed hot water heating application.

Figure 12.2 shows a photograph and technical description of the DT2811 multifunction board. Like other I/O boards there are a variety of accessories available for this product. Screw terminal boxes similar to that shown in Figure 5.10 are available for this product also cables and thermocouple cold-junction conditioning panels.

Contec ADC-100 Series

Many multifunction boards, produced by a variety of manufacturers, are available on the market. Most producers of interfacing products offer both the inexpensive single function boards, and the flexible multifunction boards. As expected each manufacturer offers a selection of multifunction boards, each with different combinations of functions and capabilities. An example of another multifunction board with a different combination of functions is the ADC-100 produced by Contec Microelectronics U.S.A. Inc.

The ADC-100 Series board, shown in Figure 12.3, offers 16 single-ended analog input channels, eight differential input channels, and two analog output channels. This board differs from the previously described multifunction board, because it also offers 24 TTL I/O lines. The 24 digital lines are interfaced through an 8255 PPI chip; therefore, the various I/O combinations of ports A, B, and C, described in Chapter 5, are available to the user of this board.

An interesting aspect of any multifunction board is that while the board uses fewer slots in the computer's mother board than separate single function boards, the multifunction boards do require several consecutive I/O addresses. Recall that in Chapter 5 the suggested port address for the PIO-12 was 300 Hex or another

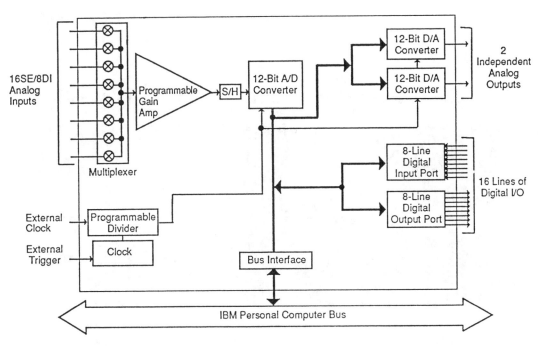

Figure 12.1 DT2811 Multifunction I/O Board Block Diagram *(Courtesy Data Translation, Inc)*

address within the range 300 Hex to 31F Hex. A single PIO-12 board required four adjacent address locations, one address for the control register and an address for each port identified as A, B, and C. By comparison the ADC-100 multifunction board requires 64 consecutive address locations. This requirement is due to the number of different functions and registers that this board contains. This number of consecutive addresses is not unusual for a multifunction board. However, it is a factor that must be considered when selecting port addresses for one or more multifunction boards or other devices requiring port addresses.

Keithley MetraByte DAS-HRES

A third multifunction board option is the DAS-HRES, shown in Figure 12.4. This board is produced by Keithley MetraByte Corporation. The DAS-HRES is a high-performance multifunction board that offers 16-bit resolution with eight fully differential input analog channels. This board combines very high resolution, gain control, analog output, and both digital input and digital output.

The 16-bit resolution of the analog-to-digital converter and the high-speed characteristic of the A/D conversion make this board different from other boards. The DAS-HRES uses a high performance sampling A/D converter with a conversion time of 17.25 microseconds. This conversion time provides for a maximum

IBM PC

DATA TRANSLATION®

DT2811
Low Cost Analog and Digital I/O Board For PC/XT/AT Compatibles

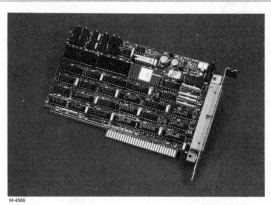

M-4566

The DT2811 is a low cost multifunction board for IBM® PC/XT/AT compatibles. It provides 12-bit A/D resolution with 20kHz throughput. A pacer clock is provided to control A/D sampling and interrupts are supported to signal a host program of important board events. Two models are available—one with gains of 1, 2, 4, or 8 and one with gains of 1, 10, 100, or 500. The DT2811 is supported by the LPCLAB™ Subroutine Library and many application software packages.

Low Cost, Multifunction Board
The DT2811 gives you 12-bit 20kHz A/D, 2 D/A channels, 16 lines of digital I/O, and a programmable pacer clock—all at low cost on a single IBM PC/XT/AT compatible board.

Measure 5mV Full Scale
Sampling of very low level signals (to 5mV), such as thermocouple outputs, with no loss of A/D resolution is possible using programmable gains to 500.

Pacer Clock Controls A/D Sampling
An onboard programmable pacer clock lets you automatically perform repeated A/D conversions at the maximum speed allowed by the board—20kHz.

Two DACs, 16 DIO Lines
Two D/A converters with 12-bit resolution and 50kHz throughput each and 16 digital I/O (DIO) lines are provided for control and measurement applications, where output lines are needed to alter process or experiment parameters based on sampled data.

Interrupt-Driven Data Acquisition
Support for interrupt on A/D conversion done or error simplifies programming and frees the host CPU to perform other operations.

Wide Variety of Software Support
Included free of charge with the DT2811 are example programs in BASIC that demonstrate all board functions and menu-driven DT/Gallery™ for calibration, data logging, and strip chart recording. The LPCLAB™ Subroutine Library, as well as several application software packages including LABTECH NOTEBOOK® for Windows and GENESIS® also support the DT2811.

Summary

A/D: 12 bits; 20kHz throughput; 16SE/8DI; gains to 500

D/A: 2 DACs; 12 bits; 50kHz throughput per DAC

Digital I/O: 8 lines in, 8 lines out

Clocks: One programmable clock (controls A/D operations)

Interface: Programmed I/O; one interrupt

Figure 12.2 DT2811 Multifunction I/O Board *(Courtesy Data Translation, Inc)*

ADC-100 Series

*Multifunction
A/D, D/A, DIO
ADC-100, ADC-200
ADC-300*

Features:

ADC-100

- *Plug-in board for IBM PC/XT/AT Bus*

- *16 single-ended/8 differential analog input channels*

- *12-bit A/D, 50,000 samples/sec. with DMA*

- *2 analog output channels, 12-bit D/A, 30,000 samples/sec.*

- *24 programmable digital I/O*

- *Interrupt handling*

- *Programmable scan rate*

- *Foreground/background operation*

- *Easy-to-use software support with sample program*

- *High-level utility program support: Labtech Notebook, Labtech Acquire, Labtech Control, OnSpec, UnkelScope, SnapShot, Module-PAC,*

- *Real time clock/calendar with battery back-up*

ADC-200 (ADC-100 plus)

- *Analog inputs/outputs can be operated in current loop mode*

ADC-300 (ADC-100 plus)

- *Programmable voltage gain selection*

Specifications

- ■ Analog I/O

- ■ Input

- Analog Input:
  16 Single ended/8 Differential channels

- Full Scale Input Range:
  Bipolar ±5V, ±10V;
  Unipolar 0 – 10V;
  Current loop (ADC-200 only) 4..20mA

- Absolute maximum input voltage ±30V

- Input impedance: > 1 megohm

- Programmable gain (ADC-300 only): 1, 10, 100, 200

- A/D Conversion:
  successive approximation, 12-bit resolution, throughput 50,000 samples/second

- Accuracy: ±0.04% of FSR at 25°C

- Zero drift: ±20 ppm of FSR per °C

- Gain Drift: ±50 ppm of FSR per °C

- Differential Linearity Drift:
  ±3 ppm of FSR per °C

- ■ Output

- Analog outputs: 2

- Output Range:
  Bipolar ±2.5V, ±5V, ±10V,
  Unipolar 0 to 5V, 0 to 10V
  Current loop (ADC-200 only) 4..20mA

- Output current: ±5 mA

- Output impedance (DC):
  0.2 ohm max.

- Capacitive drive capability:
  0.5 microfarad

- D/A conversion:
  ladder resistor network, 12-bit resolution, throughput 30,000 samples/second

- Accuracy: ±0.05% of FSR at 25°C

- Slew Rate: 10V/microsecond

- Settling Time to 1/2 LSB:
  4 microsecond, 20V step

- ■ Digital I/O

- Number of channels: 24

- I/O Logic: TTL

- I/O Interface: 8255A

- Input load: 1 LS-TTL

- Fan out: 2 LS-TTL

Figure 12.3 ADC-100 Series Photograph and Specification Sheet
(Courtesy Contec Microelectronics U.S.A. Inc)

**47 KILOSAMPLE/SEC, 16-BIT
ANALOG AND DIGITAL I/O BOARD**

DAS-HRES

FEATURES

- 16-bit resolution
- 8 fully differential input channels
- 2 channels of 16-bit analog output
- 3 counter/timers
- Software-selectable gains 1, 2, 4 and 8
- 16 bits digital I/0
- 47 kS/s acquisition rate
- Unipolar and bipolar operation

APPLICATIONS

- Signal analysis
- Sensor interface
- Chromatography
- Process monitor/control
- Laboratory automation

BLOCK DIAGRAM

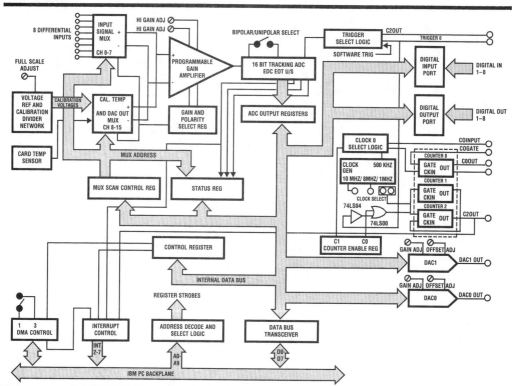

Figure 12.4 DAS-HRES Photograph and Data Sheet *(Courtesy Keithley MetraByte Corporation)*

sampling rate of 47,600 samples per second. The combination of 16-bit resolution and quick conversion time places this board in a different league from entry level DAC-4 A/D product offered by the same vendor.

Another feature of the Keithley MetraByte DAS-HRES product is the flexibility of gain control. Through the gain control and the resolution of the 16-bit converter the resolution of the board rises to 19 uV per digital bit. The DAS-HRES is a "high end" product that offers very fine A/D resolution, fast conversion, and multifunction features on a single board.

Multifunction boards are definitely an option to consider when purchasing I/O boards. The decision to buy a multifunction board depends upon several factors including the availability of slots in the computer, the number and variety of I/O functions needed for an application, and the cost of the multifunction board. If the purchase of the sophisticated multifunction board means a higher cost and results in unused functions, then the wisdom of investing money in untapped functions is questionable. On the other hand, a multifunction board may be a more economical investment than three or four separate I/O boards.

Most vendors of interfacing products offer a selection of multifunction boards and accessories. A review of available products and a comparison of specifications and products from different vendors can be a good learning experience.

REAL-TIME DATA ACQUISITION/ANALYSIS SOFTWARE

The second major trend that is influencing the field of data acquisition and control is the popularity of commercially available data acquisition and analysis software. These software packages offer engineers or technicians the option of using menu-driven software to control the gathering of data as well as the analysis of data without the need to write a unique software program. These packages do not eliminate the need to write programs as was done in earlier chapters in both the C language and BASIC. Writing software to control the input of data, manipulation of the data by the computer, and the output of signals to control external devices continues to be accomplished by writing software, however, other options do exist for engineers and technicians.

Commercially available packages do offer many attractive features including the ability to integrate pre-written software with menu selection, the ability to integrate the software with a variety of interfacing boards, and the ability of the software to graphically display data that has been input to the computer. The data can be displayed in real-time, as the data is read by the computer, or the data can be displayed later after it has been logged to memory or disk drive. In either case commercial software offers the user the flexibility of filtering the data through the inclusion of high-pass, low-pass, or band-pass software filters.

Figure 12.5 shows a typical setup menu screen for a data acquisition software product. In this example the software product is Labtech Notebook, produced by Labtech, 400 Research Drive, Wilmington, MA 01887.

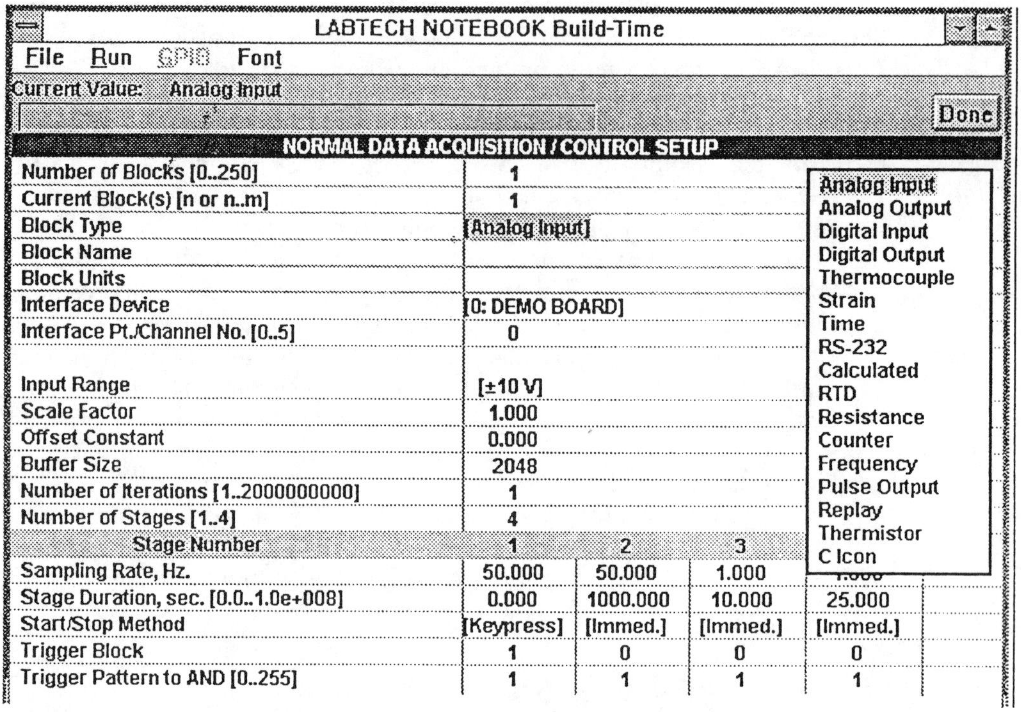

```
LABTECH NOTEBOOK Build-Time
File   Run   GPIB   Font
Current Value:   Analog Input
                                                              [Done]
```

NORMAL DATA ACQUISITION / CONTROL SETUP

Number of Blocks [0..250]	1			Analog Input
Current Block(s) [n or n..m]	1			Analog Output
Block Type	[Analog Input]			Digital Input
Block Name				Digital Output
Block Units				Thermocouple
Interface Device	[0: DEMO BOARD]			Strain
Interface Pt./Channel No. [0..5]	0			Time
				RS-232
Input Range	[±10 V]			Calculated / RTD
Scale Factor	1.000			Resistance
Offset Constant	0.000			Counter
Buffer Size	2048			Frequency
Number of Iterations [1..2000000000]	1			Pulse Output
Number of Stages [1..4]	4			Replay

Stage Number	1	2	3	Thermistor / C Icon
Sampling Rate, Hz.	50.000	50.000	1.000	1.000
Stage Duration, sec. [0.0..1.0e+008]	0.000	1000.000	10.000	25.000
Start/Stop Method	[Keypress]	[Immed.]	[Immed.]	[Immed.]
Trigger Block	1	0	0	0
Trigger Pattern to AND [0..255]	1	1	1	1

Figure 12.5 Menu Screen for Labtech Notebook *(Courtesy Laboratory Technologies Corporation)*

The menu offers the user the opportunity to define a variety of parameters including the number of input channels being used, the type of channel, the unit of measure for the specified channel, along with a variety of other parameters including the sampling rate. Setup menus of this type are used to set the behavior characteristics of each of four data acquisition functions. The functions include *data input*, *data output*, *data storage*, and *data display*. In the example shown in Figure 12.5 an analog input channel is being specified. This input is identified as a voltage signal with a voltage range of ±10 volts. The Sampling Rate for this example is 50 readings per second.

Setup information is specifically adapted for the interfacing board used in a particular application. Most data acquisition software includes I/O drivers for most popular interfacing boards including both analog and digital boards. Sometimes, in an effort to keep costs low, producers of the software require that the purchaser of the software specify which I/O boards or group of boards that the software will be used with. This specificity reduces the cost by limiting the number of drivers included in the software. Producers of data acquisition software advertise that their products are compatible with upwards of 400 different interfacing boards.

Besides the major tasks of inputting data and storing data the software is capable of displaying the data in a graphical format. Figure 12.6 shows a picture of data readings being displayed on a CRT screen along with a second CRT trace showing the smoothed out data readings after the raw data has been filtered. This figure is provided by Unkel Software Inc., 442 Marrett Road, Lexington, MA 02173. The figure is a CRT print that was created by a product produced by Unkel called UnkelScope Junior. In the upper portion of the figure the raw data is illustrated by many data points. Each point represents a unique data reading. The lower portion of the figure shows the data curve after filtering. Here the data has been smoothed by filtering the raw data readings through a low pass filter.

Smoothing raw data can be a powerful addition to data acquisition software as it reduces a lot of confusing individual data points into a clearly identifiable trend. The Unkel software gives the user the option of displaying up to four input traces on a single screen along with specifying trace colors.

The following features and specifications, shown in Figure 12.7, are applicable to most commercially available entry level data acquisition software. More

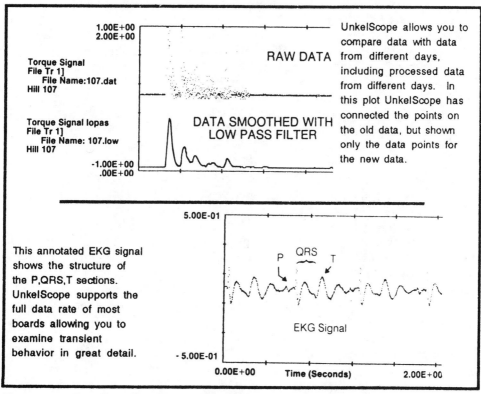

Figure 12.6 Raw Data Display and Filtered Data Display *(Courtesy Unkel Software Inc)*

— **Completely menu driven.**
— Programming Required
— See data in real time or as quickly as hardware allows
— Display measured signals vs. time or vs. other measured signals
— Zoom in on data. High resolution plots
— Read out of time and voltage from two independent cursors; other graphics data manipulations
— Store, retrieve data files; labels, time, date stored for the record
— Compare current data with stored data
— Linear Conversion to show measurements in physical units
— Hardcopy plots direct from program
— Extensive triggering options

Figure 12.7 Data Acquisition Software Specifications *(Courtesy Unkel Software Inc)*

advanced software packages provide other features including compatibility with Microsoft Windows, the use of icons, and object-orientated graphics.

Figure 12.8 shows a photograph of a CRT screen in which the data acquisition software is operating under Windows. In this example, Labtech Notebook is operating together with Microsoft Windows to create a multitasking environment. The user can collect data, while simultaneously display data, logging data to disk, and making control decisions, all while using other Windows applications.

Another feature of software packages is the inclusion of the ability to provide object-oriented graphics as part of the CRT output. Object-orientated graphics allow the application engineer to include representations of various systems and objects on the screen as part of the data acquisition and control process. Typical of the graphics that might be included on a screen are machinery items such as simulated conveyor belts, robot arms, and other mechanical items.

The sketch in Figure 12.9 shows two conveyor belts with material flowing on the belts. This type of picture is ideal for a graphic display because it shows a process taking place. If this picture were put into graphic form the conveyors would show motion and perhaps show material falling from belt #1 through the hopper to belt #2. If the actual conveyor belts were to stop moving the graph dispaly would show the the stationary belts. This type of display actually tells the the viewer of the CRT display a lot of information about whether the belts are moving, the level of material on the belt, and whether the material being transported is jammed.

Another application of graphics illustration that would be informative to the viewer is a graphic of a water tank in which the level of the water rises and falls as determined by an analog sensor in the system being monitored. As the water level in the actual tank being monitored rises or falls the graphic display echo's this movement of the water level. Other examples of graphics include meter faces which translate data read by an I/O board into a graphically moving a meter pointer.

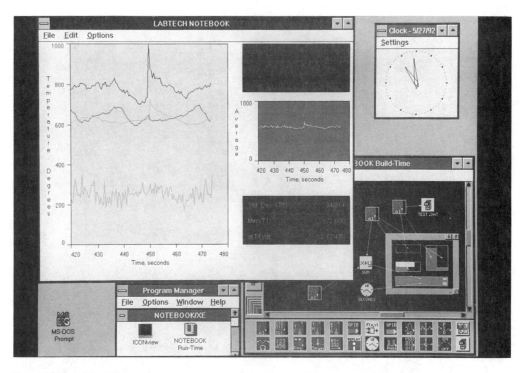

Figure 12.8 Real-Time Multitasking Under Windows *(Courtesy Laboratory Technologies Corporation)*

Figure 12.10 shows a screen print of an object-orientated graphical display. In this display, the hypothetical system being monitored by a computer is the testing of an automotive transmission. The analytical data being read by the computer is simultaneously being logged to disk while it is being graphically displayed in an interesting and user friendly format. In Figure 12.10 the information being displayed includes crankcase oil temperature, the torque output from the transmission, and the RPMs of the drive shaft. This information is displayed in an object oriented form which is both informative and visually appealing. Software of this type allows designers to create their own custom switches, knobs, and display objects.

CONCLUSION

As the reader of this text looks back over the 12 chapters, it is the authors' hope that the book has benefitted the reader. If this text has been successful it has guided the reader through the under lying concepts, the terminology, and the essential software programming commands that will allow the reader to use I/O boards in

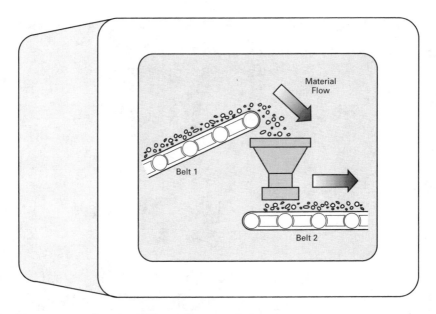

Figure 12.9 Sketch of Object Orientated Graphic

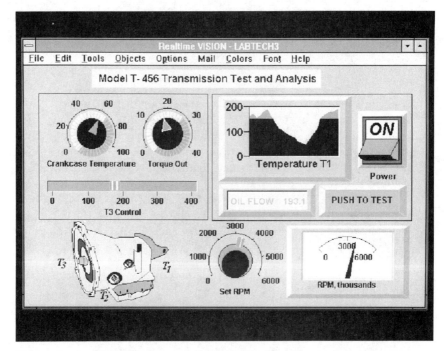

Figure 12.10 Object Orientated Graphic Display *(Courtesy Laboratory Technologies Corporation)*

a practical way. The authors have intended to introduce the reader to some common interfacing boards on the market and to the companies that produce these products.

Hopefully, there is little doubt in the readers mind that the topical area of data acquisition and control is an important body of knowledge for today's engineers and technicians. As anyone familiar with electronics, computers, and programming already knows, learning never stops and it is hoped that this text will spark further learning.

Good luck as you continue learning about the fascinating field of data acquisition and control.

BIOS Interrupt Functions

Interrupt #	Vector Address	Function
0	00	Divide by 0
1	04	Single Step
2	08	NMI
3	0C	Break Point
4	10	Overflow
5	14	Print Screen
6	18	Reserved
7	1C	Reserved
8	20	Timer - TIC
9	24	Keyboard (internal)
A	28	AT Cascaded Interrupts
B	2C	COM #2
C	30	COM #1
D	34	Line Printer #2 (internal)
E	38	Floppy Disk (internal)
F	3C	Line Printer #1 (internal)
10	40	Video Routines
11	44	System Hardware Configuration
12	48	System Memory Size
13	4C	Floppy/Fixed Disk Access
14	50	Serial Communication
15	54	Special System Extensions
16	58	Keyboard Routines
17	5C	Printer Routines
18	60	Reserved
19	64	Boot
1A	68	Timer
1B	6C	Keyboard Extension (#9)
1C	70	Timer/TIC Extension (#8)
1D	74	Video Parameter Table
1E	78	Floppy Parameter Table
1F	7C	Graphic Character Table
42	104	EGA BIOS Redirect
43	108	EGA Parameter Table
44	10C	EGA Character Table
70	1C0	AT Real-Time Clock
71	1C4	Interrupt Cascade Redirect
72	1C8	Reserved
73	1CC	Reserved
74	1D0	Mouse
75	1D4	Math Co-Processor
76	1D8	Fixed Disk

APPENDIX B

National Semiconductor

May 1990

LM34/LM34A/LM34C/LM34CA/LM34D
Precision Fahrenheit Temperature Sensors

General Description

The LM34 series are precision integrated-circuit temperature sensors, whose output voltage is linearly proportional to the Fahrenheit temperature. The LM34 thus has an advantage over linear temperature sensors calibrated in degrees Kelvin, as the user is not required to subtract a large constant voltage from its output to obtain convenient Fahrenheit scaling. The LM34 does not require any external calibration or trimming to provide typical accuracies of $\pm \frac{1}{2}°F$ at room temperature and $\pm 1\frac{1}{2}°F$ over a full -50 to $+300°F$ temperature range. Low cost is assured by trimming and calibration at the wafer level. The LM34's low output impedance, linear output, and precise inherent calibration make interfacing to readout or control circuitry especially easy. It can be used with single power supplies or with plus and minus supplies. As it draws only 75 μA from its supply, it has very low self-heating, less than 0.2°F in still air. The LM34 is rated to operate over a $-50°$ to $+300°F$ temperature range, while the LM34C is rated for a $-40°$ to $+230°F$ range ($0°F$ with improved accuracy). The LM34 series is available packaged in hermetic TO-46 transistor packages,

while the LM34C, LM34CA and LM34D are also available in the plastic TO-92 transistor package. The LM34D is also available in an 8-lead surface mount small outline package. The LM34 is a complement to the LM35 (Centigrade) temperature sensor.

Features

- Calibrated directly in degrees Fahrenheit
- Linear + 10.0 mV/°F scale factor
- 1.0°F accuracy guaranteed (at +77°F)
- Rated for full $-50°$ to $+300°F$ range
- Suitable for remote applications
- Low cost due to wafer-level trimming
- Operates from 5 to 30 volts
- Less than 90 μA current drain
- Low self-heating, 0.18°F in still air
- Nonlinearity only $\pm 0.5°F$ typical
- Low-impedance output, 0.4Ω for 1 mA load

Connection Diagrams

TO-46
Metal Can Package*

TL/H/6685-1
*Case is connected to negative pin (GND).
Order Numbers LM34H, LM34AH,
LM34CH, LM34CAH or LM34DH
See NS Package Number H03H

TO-92
Plastic Package

BOTTOM VIEW
TL/H/6685-2
Order Number LM34CZ,
LM34CAZ or LM34DZ
See NS Package Number Z03A

SO-8
Small Outline Molded Package

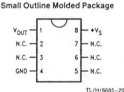

TL/H/6685-20
Top View
N.C. = No Connection
Order Number LM34DM
See NS Package Number M08A

Typical Applications

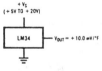

TL/H/6685-3
FIGURE 1. Basic Fahrenheit Temperature Sensor
($+5°$ to $+300°F$)

CHOOSE $R_1 = (-V_S)/50 \ \mu A$
$V_{OUT} = +3,000$ mV AT $+300°F$
$= +750$ mV AT $+75°F$
$= -500$ mV AT $-50°F$

TL/H/6685-4
FIGURE 2. Full-Range Fahrenheit Temperature Sensor

Absolute Maximum Ratings (Note 10).

If Military/Aerospace specified devices are required, please contact the National Semiconductor Sales Office/Distributors for availability and specifications.

Supply Voltage	+35V to −0.2V
Output Voltage	+6V to −1.0V
Output Current	10 mA

Storage Temperature,

TO-46 Package	−76°F to +356°F
TO-92 Package	−76°F to +300°F
SO-8 Package	−65°C to +150°C
ESD Susceptibility (Note 11)	800V

Lead Temp.

TO-46 Package (Soldering, 10 seconds)	+300°C
TO-92 Package (Soldering, 10 seconds)	+260°C

SO Package (Note 12):

Vapor Phase (60 seconds)	215°C
Infrared (15 seconds)	220°C

Specified Operating Temp. Range (Note 2)

	T_{MIN} to T_{MAX}
LM34, LM34A	−50°F to +300°F
LM34C, LM34CA	−40°F to +230°F
LM34D	+32°F to +212°F

DC Electrical Characteristics (Note 1, Note 6)

Parameter	Conditions	LM34A			LM34CA			Units (Max)
		Typical	Tested Limit (Note 4)	Design Limit (Note 5)	Typical	Tested Limit (Note 4)	Design Limit (Note 5)	
Accuracy (Note 7)	$T_A = +77°F$	±0.4	±1.0		±0.4	±1.0		°F
	$T_A = 0°F$	±0.6			±0.6		±2.0	°F
	$T_A = T_{MAX}$	±0.8	±2.0		±0.8	±2.0		°F
	$T_A = T_{MIN}$	±0.8	±2.0		±0.8		±3.0	°F
Nonlinearity (Note 8)	$T_{MIN} \le T_A \le T_{MAX}$	±0.35		±0.7	±0.30		±0.6	°F
Sensor Gain (Average Slope)	$T_{MIN} \le T_A \le T_{MAX}$	+10.0	+9.9, +10.1		+10.0	+9.9, +10.1		mV/°F, min mV/°F, max
Load Regulation (Note 3)	$T_A = +77°F$	±0.4	±1.0		±0.4	±1.0		mV/mA
	$T_{MIN} \le T_A \le T_{MAX}$ $0 \le I_L \le 1$ mA	±0.5		±3.0	±0.5		±3.0	mV/mA
Line Regulation (Note 3)	$T_A = +77°F$	±0.01	±0.05		±0.01	±0.05		mV/V
	$5V \le V_S \le 30V$	±0.02		±0.1	±0.02		±0.1	mV/V
Quiescent Current (Note 9)	$V_S = +5V, +77°F$	75	90		75	90		μA
	$V_S = +5V$	131		160	116		139	μA
	$V_S = +30V, +77°F$	76	92		76	92		μA
	$V_S = +30V$	132		163	117		142	μA
Change of Quiescent Current (Note 3)	$4V \le V_S \le 30V, +77°F$	+0.5	2.0		0.5	2.0		μA
	$5V \le V_S \le 30V$	+1.0		3.0	1.0		3.0	μA
Temperature Coefficient of Quiescent Current		+0.30		+0.5	+0.30		+0.5	μA/°F
Minimum Temperature for Rated Accuracy	In circuit of *Figure 1*, $I_L = 0$	+3.0		+5.0	+3.0		+5.0	°F
Long-Term Stability	$T_j = T_{MAX}$ for 1000 hours	±0.16			±0.16			°F

Note 1: Unless otherwise noted, these specifications apply: −50°F ≤ T_j ≤ + 300°F for the LM34 and LM34A; −40°F ≤ T_j ≤ +230°F for the LM34C and LM34CA; and +32°F ≤ T_j ≤ + 212°F for the LM34D. V_S = + 5 Vdc and I_{LOAD} = 50 μA in the circuit of *Figure 2*; + 6 Vdc for LM34 and LM34A for 230°F ≤ T_j ≤ 300°F. These specifications also apply from + 5°F to T_{MAX} in the circuit of *Figure 1*.

Note 2: Thermal resistance of the TO-46 package is 720°F/W junction to ambient and 43°F/W junction to case. Thermal resistance of the TO-92 package is 324°F/W junction to ambient. Thermal resistance of the small outline molded package is 400°F/W junction to ambient. For additional thermal resistance information see table in the Typical Applications section.

Note 3: Regulation is measured at constant junction temperature using pulse testing with a low duty cycle Changes in output due to heating effects can be computed by multiplying the internal dissipation by the thermal resistance.

Note 4: Tested limits are guaranteed and 100% tested in production.

Note 5: Design limits are guaranteed (but not 100% production tested) over the indicated temperature and supply voltage ranges. These limits are not used to calculate outgoing quality levels.

Note 6: Specification in BOLDFACE TYPE apply over the full rated temperature range.

Note 7: Accuracy is defined as the error between the output voltage and 10 mV/°F times the device's case temperature at specified conditions of voltage, current, and temperature (expressed in °F).

Note 8: Nonlinearity is defined as the deviation of the output-voltage-versus-temperature curve from the best-fit straight line over the device's rated temperature range.

Note 9: Quiescent current is defined in the circuit of *Figure 1*.

Note 10: Absolute Maximum Ratings indicate limits beyond which damage to the device may occur. DC and AC electrical specifications do not apply when operating the device beyond its rated operating conditions (see Note 1).

Note 11: Human body model, 100 pF discharged through a 1.5 kΩ resistor.

Note 12: See AN-450 "Surface Mounting Methods and Their Effect on Product Reliability" or the section titled "Surface Mount" found in a current National Semiconductor Linear Data Book for other methods of soldering surface mount devices.

Typical Applications

The LM34 can be applied easily in the same way as other integrated-circuit temperature sensors. It can be glued or cemented to a surface and its temperature will be within about 0.02°F of the surface temperature. This presumes that the ambient air temperature is almost the same as the surface temperature; if the air temperature were much higher or lower than the surface temperature, the actual temperature of the LM34 die would be at an intermediate temperature between the surface temperature and the air temperature. This is expecially true for the TO-92 plastic package, where the copper leads are the principal thermal path to carry heat into the device, so its temperature might be closer to the air temperature than to the surface temperature.

To minimize this problem, be sure that the wiring to the LM34, as it leaves the device, is held at the same temperature as the surface of interest. The easiest way to do this is to cover up these wires with a bead of epoxy which will insure that the leads and wires are all at the same temperature as the surface, and that the LM34 die's temperature will not be affected by the air temperature.

The TO-46 metal package can also be soldered to a metal surface or pipe without damage. Of course in that case, the V_ terminal of the circuit will be grounded to that metal. Alternatively, the LM34 can be mounted inside a sealed-end metal tube, and can then be dipped into a bath or screwed into a threaded hole in a tank. As with any IC, the LM34 and accompanying wiring and circuits must be kept insulated and dry, to avoid leakage and corrosion. This is especially true if the circuit may operate at cold temperatures where condensation can occur. Printed-circuit coatings and varnishes such as Humiseal and epoxy paints or dips are often

used to insure that moisture cannot corrode the LM34 or its connections.

These devices are sometimes soldered to a small, lightweight heat fin to decrease the thermal time constant and speed up the response in slowly-moving air. On the other hand, a small thermal mass may be added to the sensor to give the steadiest reading despite small deviations in the air temperature.

Capacitive Loads

Like most micropower circuits, the LM34 has a limited ability to drive heavy capacitive loads. The LM34 by itself is able to drive 50 pF without special precautions. If heavier loads are anticipated, it is easy to isolate or decouple the load with a resistor; see *Figure 3*. Or you can improve the tolerance of capacitance with a series R-C damper from output to ground; see *Figure 4*. When the LM34 is applied with a 499Ω load resistor (as shown), it is relatively immune to wiring capacitance because the capacitance forms a bypass from ground to input, not on the output. However, as with any linear circuit connected to wires in a hostile environment, its performance can be affected adversely by intense electromagnetic sources such as relays, radio transmitters, motors with arcing brushes, SCR's transients, etc., as its wiring can act as a receiving antenna and its internal junctions can act as rectifiers. For best results in such cases, a bypass capacitor from V_{IN} to ground and a series R-C damper such as 75Ω in series with 0.2 or 1 μF from output to ground are often useful. These are shown in the following circuits.

Temperature Sensor,
Single Supply, −50° to +300°F

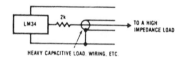

TL/H/6685-6

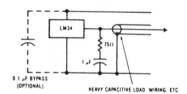

FIGURE 3. LM34 with Decoupling from Capacitive Load

FIGURE 4. LM34 with R-C Damper

Temperature Rise of LM34 Due to Self-Heating (Thermal Resistance)

Conditions	TO-46, No Heat Sink	TO-46, Small Heat Fin*	TO-92, No Heat Sink	TO-92, Small Heat Fin**	SO-8 No Heat Sink	SO-8 Small Heat Fin**
Still air	720°F/W	180°F/W	324°F/W	252°F/W	400°F/W	200°F/W
Moving air	180°F/W	72°F/W	162°F/W	126°F/W	190°F/W	160°F/W
Still oil	180°F/W	72°F/W	162°F/W	126°F/W		
Stirred oil	90°F/W	54°F/W	81°F/W	72°F/W		
(Clamped to metal, infinite heat sink)	(43°F/W)				(95°F/W)	

*Wakefield type 201 or 1" disc of 0.020" sheet brass, soldered to case, or similar.

**TO-92 and SO-8 packages glued and leads soldered to 1" square of 1/16" printed circuit board with 2 oz copper foil, or similar.

Typical Applications (Continued)

Two-Wire Remote Temperature Sensor
(Grounded Sensor)

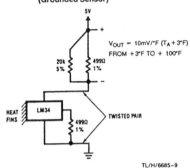

$$V_{OUT} = 10mV/°F \ (T_A + 3°F)$$
$$FROM +3°F \ TO \ +100°F$$

TL/H/6685–9

Two-Wire Remote Temperature Sensor
(Output Referred to Ground)

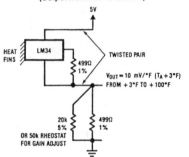

$$V_{OUT} = 10 \ mV/°F \ (T_A + 3°F)$$
$$FROM +3°F \ TO \ +100°F$$

TL/H/6685–10

4-to-20 mA Current Source
(0 to +100°F)

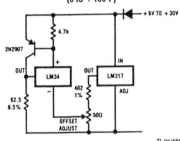

TL/H/6685–11

Fahrenheit Thermometer
(Analog Meter)

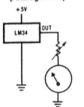

TL/H/6685–12

Expanded Scale Thermometer
(50° to 80° Fahrenheit, for Example Shown)

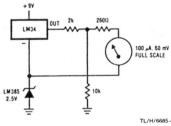

100 µA, 50 mV
FULL SCALE

TL/H/6685–13

Temperature-to-Digital Converter
(Serial Output, +128°F Full Scale)

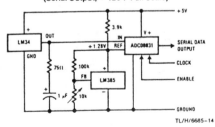

TL/H/6685–14

Pressure Sensor Information

 **SenSym**

142SC Series
0–1psi to 0–150 psi
Signal Conditioned
Pressure Transducers

FEATURES

■ Improved Performance Replacement for Honeywell/Microswitch 140PC Series

■ High Level Voltage Output

■ Field Interchangeable

■ Calibrated and Temperature Compensated

APPLICATIONS

■ Medical Equipment

■ Barometry

■ Computer Peripherals

■ HVAC

GENERAL DESCRIPTION

The 142SC series transducers provide a 1–6V output which is directly proportional to applied pressure. This series consists of eight (8) devices for monitoring differential, gage, or absolute pressures from 0–1 to 0–150 psi. These products feature a high level voltage output, complete calibration and temperature compensation.

Based on Sensym's precision SX series sensors, the 142SC series is an improved performance, direct replacement for the Honeywell/Microswitch 142PC series with equivalent pinout and package mounting dimensions.

This allows direct replacement in existing PC board layouts for the Microswitch parts. Sensyms 142SC devices offer the added advantage of tighter tolerances which give greater accuracy and field interchangeability.

These products are designed to be used with non-corrosive, non-ionic gases and liquids. For more demanding or corrosive media applications, Sensym's ST2000 stainless steel isolated family should be used.

FUNCTIONAL SPECIFICATIONS

142SC Series

Maximum Ratings

Supply Voltage	$+7V_{DC}$ to $16V_{DC}$
Output Current	
Source	10 mA
Sink	5 mA
Temperature Ranges	
Compensated	$-18°C$ to $+63°C$
Operating	$-40°C$ to $+85°C$
Storage	$-55°C$ to $+125°C$

Reference Conditions

Supply Voltage	$8.0 \pm 0.01 V_{DC}$
Reference Temperature	25°C
Common-mode Pressure	0 psig

INDIVIDUAL OPERATING CHARACTERISTICS

Sensym Part #	Operating Pressure Range	Proof Pressure	Sensitivity
142SC01D	0–1 psid (g)	20 psig	5 V/psi
142SC05D	0–5 psid (g)	20 psig	1 V/psi
142SC15A	0–15 psia	45 psia	333 mV/psi
142SC15D	0–15 psid (g)	45 psig	333 mV/psi
142SC30A	0–30 psia	60 psia	167 mV/psi
142SC30D	0–30 psid (g)	60 psid	167 mV/psi
142SC100D	0–100 psid (g)	200 psid	50 mV/psi
142SC150D	0–150 psid (g)	200 psid	33 mV/psi

PERFORMANCE SPECIFICATIONS (For All Devices) (Note 1)

Parameter	Min.	Typ.	Max.	Unit
Offset Calibration (Note 2)	0.95	1.0	1.05	V
Output at Full Pressure	5.90	6.0	6.10	V
Full-scale Span (Note 3)	4.95	5.0	5.05	V
Linearity ($P_2 > P_1$)	—	0.5	1.5	%FSO
($P_2 < P_1$) (Note 4)	—	0.2	0.75	%FSO
Temperature Shift ($-18°C$ to $+63°C$) (Note 5)	—	0.5	1.0	%FSO
Repeatability and Hysteresis	—	0.2	—	%FSO
Response Time	—	0.1	1.0	ms

Specification Notes:

Note 1: Performance specifications shown are at reference conditions. Specifications apply for absolute pressure devices with pressure applied to Port 1. For gage devices pressure is applied to Port 2 and Port 1 is left open to ambient. For differential pressures, Port 2 is the high pressure port. For operation at other than $8.0V_{DC}$ the typical ratiometricity error at 7 to 8V or 8 to 9V is $\pm 0.50\%$ FSO and at 9 to 12V it is $\pm 2.00\%$ FSO. All Sensym differential devices feature dual pressure ports and can be used as gage or differential sensors. For absolute devices, Port 2 is inactive.

Note 2: Offset calibration is at the lowest pressure for each given device.

Note 3: Full-scale span is the algebraic difference between the output voltage at full-scale pressure and the output at the lowest operating pressure.

Note 4: Linearity refers to the best straight line fit as measured for offset, full-scale and 1/2 full-scale pressure.

Note 5: Temperature shift refers to the combined effects of offset and sensitivity shifts. This is tested at $-18°C$ to $+63°C$ relative to 25°C. The maximum temperature shift specification applies to all devices except the 142SC01D devices which have a maximum shift of 1.5% FSO from 5°C to 45°C.

GENERAL DISCUSSION

Sensym's 142SC series utilizes Sensym's proven SX series sensor element in combination with a custom individually laser trimmed thick film ceramic. Each device is calibrated for offset and sensitivity as well as temperature effects providing an accurate, reliable sensor for a wide variety of sensor applications.

Output Characteristics

The 142SC products give an output voltage which is directly proportional to applied pressure. For the 142SC gage and differential devices, an increasing or positive going output signal will result when increasing pressure is applied to port P2. (For absolute pressure, increases in pressure applied to port P1 produce an increasing output signal. Port P2 is inactive on absolute devices.) For standard 142SC devices the output is ratiometric to the supply voltage. Changes in the supply voltage will cause proportional changes in the offset voltage and full scale span.

User Calibration

The 142SC devices are fully calibrated for offset and span and should therefore require little or no user adjustment in most applications.

Vacuum Reference (Absolute Devices)

Absolute sensors have a hermetically sealed vacuum reference chamber within the sensor chip. The offset voltage on these units is therefore measured at vacuum, 0psia. Since all pressure is measured relative to a vacuum reference, all changes in barometric pressure or changes in altitude will cause changes in the device output.

142SC Series

Media Compatibility

142SC devices are compatible with most clean dry gases. Because the sensor chip circuitry is coated with a protective silicon gel, many otherwise corrosive environments can be compatible with the sensors. As shown in the physical construction diagram below, fluids must generally be compatible with nylon, aluminum, RTV, and silicon, for use with Port 2. For questions concerning media compatibility, contact the factory.

MECHANICAL AND MOUNTING CONSIDERATIONS

The 142SC package is designed for convenient pressure connection and easy PC board mounting. The package has two mounting holes allowing firm PC board connection. Mounting screws or Sensym's plastic X-mas tree clips (Part number SCXCLP) can be used for attachment. (See Application Note SSAN-25).

For pressure attachment, tygon or silicon tubing is recommended.

All versions of the 142SC sensors have two (2) tubes available for pressure connection. For absolute devices, only port P1 is active. Applying pressure through the other port will result in pressure dead-ending into the backside of the silicon sensor and the device will not give an output signal with pressure.

For gage applications, pressure should be applied to port P2. Port P1 is then the vent port which is left open to the atmosphere. For differential pressure applications, to get proper output signal polarity, port P2 should be used as the high pressure port and P1 should be used as the low pressure port.

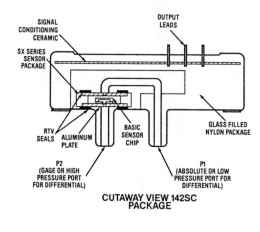

CUTAWAY VIEW 142SC PACKAGE

MOUNTING DIMENSIONS   (For Reference Only)   142SC Series

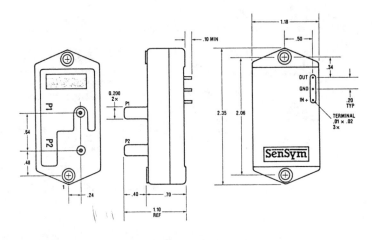

ORDERING INFORMATION   To order, use the following part numbers:

Standard Device Types

Sensym Part #	Operating Pressure Range	Honeywell/Microswitch Equivalent Part #
142SC01D	0–1 psid (g)	142PC01 (D, G)
142SC05D	0–5 psid (g)	142PC05D
142SC15A	0–15 psia	142PC15A
142SC15D	0–15 psid (g)	142PC15 (D, G)
142SC30A	0–30 psia	142PC30A
142SC30D	0–30 psid (g)	142PC30 (D, G)
142SC100D	0–100 psid (g)	N/A
142SC150D	0–150 psid (g)	N/A

Note: All Sensym differential devices feature dual pressure ports and can be used as gage or differential sensors. Sensym's differential (D) devices are therefore interchangeable with the Microswitch differential (D) or gage (G) style devices. The 141 and 143 style devices are also available for vacuum and pressure/vacuum applications. Contact the Sensym factory for details. In addition, devices which offer internal voltage regulators are also available from Sensym.

Glossary*

Accuracy—The maximum deviation that can be expected between the meter reading and the actual value being measured, under specified conditions.

A/D, D/A—Analog Input boards convert incoming analog signals into digital values (**A/D**—**A**nalog to **D**igital), allowing your PC to acquire data from analog sources; Analog Outputs convert **D/A** (**D**igital to **A**nalog), usually supplying a voltage.

A/D Conversion Time—This is the length of time the board requires to convert an analog signal into a digital value. The theoretical maximum speed (conversions/second) is the inverse of this value. The **FAST 1611** high-speed A/D board is a highly efficient example, providing a full 1 MHz speed from a 1 μs conversion time. See **Speed/Typical Throughput**.

Alias—A false lower frequency component that appears in sampled data at too low a sampling rate.

Analog—Having the property (electrical) of varying in continuous, rather than incremental or discrete steps.

Analog Slope Trigger—Sampling can be triggered at a user-selectable point on an incoming analog slope. Triggering can be set to occur at a specific threshold level, including select modes of ± slope, level high, and level low. **DAP, FAST, Flash,** and **HSDAS** series boards all include this useful triggering option.

*Courtesy CyberResearch, Inc.

Asynchronous—A method of serial communication where data is sent when it is ready without being referenced to a timing clock, rather than waiting until the receiver signals that it is ready to receive.

Attenuation—Reducing the amplitude of a waveform without introducing distortion. An adjustable passive network (filter) may be used to reduce power level of a signal without introducing any appreciable distortion.

Auto-Polarity—A characteristic of digital instruments to measure and display values of either polarity without the need to interchange test lead connections.

Autoranging—An autoranging board can be set to monitor the incoming signal and automatically select an appropriate gain level based on the previous incoming signals. **ACPC** boards are easily set to autorange for maximum flexibility in data gathering.

Base Address—A memory address that serves as the starting address for programmable registers. All other addresses are located by adding to the base address. The default base address on most of our Data Acquisition boards is 300Hex.

Baud Rate—This is the speed (in bits-per-second) at which data transfer can occur. 38.4 Kilobaud is 38,400 bits (4,800 bytes) of data per second.

Bias Current—Current that flows out of an amplifier's input terminals that will produce a voltage drop across the source impedance—in a perfect amplifier this error term will be equal to zero.

Bipolar Inputs—Bipolar inputs are designed to accept voltages in the $\pm$ X Volts range, allowing positive or negative voltage inputs. (Example: $\pm$ 5V).

Bits and Bytes—One *bit* is one binary digit, either a binary 0 or 1. One *Byte* is the amount of memory needed to store each character of information (text or numbers). There are eight *bits* to one *Byte* (or character); there are 1024 *Bytes* to one Kilobyte (K, KB), and 1024 Kilobytes to one Megabyte (MB, Meg). Our data acquisition boards typically take two-byte Samples, so a board acquiring data at a 20 KHz sample rate is actually gathering 40,000 bytes of data per second.

Buffer—A storage area for data that is used to compensate for the speed difference when transferring data from one device to another. Usually refers to an area reserved for I/O operations, into which data is read, or from which data is written.

Burst-Mode—A high speed data transfer in which the address of the data is sent followed by back-to-back data words while a physical signal is asserted.

Bus—The expansion connector built into the computer. Boards are inserted into this connector, and all communication between the computer and your board occurs through the computer's bus. There are several different expansion buses available, including the XT, AT, EISA, and Micro-Channel buses for IBM-compatible PCs, and NuBus for the Macintosh PC line.

Cache—High-speed processor memory that buffers commonly used instructions or data to increase processing throughput.

Cold Junction Compensation—Thermocouple measurements can easily be affected by the interface hardware the thermocouples are connected to. Cold Junction Compensation circuitry compensates for inaccuracies introduced in the conversion process. **STT** terminal panels feature a heavy isothermal plate for high-accuracy cold junction compensation.

Cold Junction Compensation Channel—This is an additional data acquisition input channel used exclusively for cold junction compensation, leaving all of the standard input channels free to be used for data acquisition. **ACPC** and **PC 73** boards include this channel for increased accuracy and consistency.

Common Mode Rejection Ratio (CMRR).—The board's ability to measure only the difference between the leads of a transducer, rejecting what the leads have in common. The higher the CMRR, the better the accuracy.

Conversion Rate—The number of analog to digital conversions performed per second by an A/D device.

Counter/Timer Trigger—On-board Counter/Timer circuitry can be set to trigger data acquisition at a user-selectable rate for a particular length of time.

Counter/Timers—User-accessible circuitry built into many of our DAS boards which can be used for event counting or frequency measurement.

Crest Factor—The ratio of the maximum (crest) value of a periodic function (such as AC voltage or current) to its RMS value.

Current Inputs—A board rated for current inputs can accept and convert analog current levels directly, without conversion to voltage.

Current Sink—This is the amount of current the board can supply for digital output signals. With 10–12mA or more of current sink capability, you can turn relays on and off. Digital I/O boards with less than 10–12mA of sink capability are designed for data transfer only, not power relay module switching.

DAS—Acronym for **D**ata **A**cquisition / **DA** System.

Delta-Sigma Modulating ADC—A high accuracy analog-to-digital converter circuit that samples at a higher rate and lower resolution that is needed and (by means of feedback loops) pushes the quantization noise above the frequency range of interest. This out-of-band noise is typically removed by digital filters.

Differential—See **Number of Channels**.

Direct Memory Access (DMA)—Allows a 64 KByte block of memory to be set aside for high-speed transfer of data to PC system memory. Boards which support single DMA access can store up to 64 KBytes of data (32K samples) at speeds which can reach 250 KBytes per second on XT/AT computers. Dual DMA access means that a second 64 KByte block is set up while the first is being filled, allowing unlimited sample sizes. On **EISA 2000** 1 MHz data acquisition board for EISA bus can perform data transfers at speeds of up to 16 Megabytes per second.

Drift—Meter reading or set point variations due to changes in component values, often due to external changes in ambient temperature or line voltage.

Dual-Ported RAM Memory—Allows acquired data to be transferred from on-board memory to the computer's memory while data acquisition is occurring.

Dual-Slope Conversion—A digital technique for converting a measured analog voltage or current to a precise digital equivalent. During a fixed interval of time, the output of an integrating A/D circuit rises linearly at a rate proportional to the measured analog quantity. The circuit is then switched to a precise reference voltage source of opposite polarity, causing the output to descend at a fixed rate, while an internal clock/counter generates pulses. As the output reaches its base level, the count is terminated. The total count, numerically equal to the analog input, is then output as a digital equivalent.

External Pulse Trigger—Almost all of our A/D boards allow sampling to be triggered by a voltage pulse from an external source.

FIFO—A First In First Out memory buffer; the first data received and stored is the first data sent to the acceptor.

Flash ADC—An analog to digital converter whose output code is determined in a single step by a bank of comparators and encoding logic.

Function—A set of software instructions executed by a single line of code that may have input and/or output parameters and returns a value when executed (e.g. $Y = COS(x)$).

General Purpose Interface Bus (GPIB)—The popular name for the IEEE-488 interface connection and communications standard.

Gain—Applied to the incoming signal, gain acts as a multiplication factor on the signal, increasing the number of ranges the board is designed to accept. For example, if you select the $\pm 5V$ range and set the gain to 10, signals in the $\pm.5V$ (500mV) range are usable; with a gain of 20, the range is $\pm 250mV$.

Graphical User Interface (GUI)—An intuitive, easy to use means of communicating information to and from a computer program by means of graphical screen displays. GUIs can resemble the front panels of instruments or other objects associated with a computer program.

Hertz (Hz)—A unit of frequency equal to one cycle per second. Throughout rates of 1KHz (1000Hz) imply the ability to handle 1000 samples/second, while sampling rates of 1 million samples/second are referred to as 1 Megahertz (1 MHz).

Hierarchical—A method of organizing a computer program or a system design with a series of levels; each level with further subdivisions, as in a pyramid or tree structure.

Hysteresis—An error resulting from the inability of an electrical signal to produce the same reading when approached slowly from either direction.

Icon—A graphic representation of a function or functions to be performed by the computer. For example, using a picture of a chess piece to represent a computer chess game.

Individual Gain Per Channel—Allows you to select an individual gain level for each input channel, thereby allowing a much wider range of input levels and types without sacrificing accuracy on low-level signals. Our **MIO** and **AC** series boards support individual gain settings on each channel, thereby allowing you to connect a large variety of transducers to a single board.

Integrating ADC—An analog to digital converter whose output code represents the average value of input voltage over a given time interval.

Interrupt—To stop an execution of a program, to execute another subroutine, in such a way that the first program can be resumed normally after the execution of the interrupt service program. In an IBM compatible computer there are eight levels of hardware interrupts, IRQ0, IRQ7, and 256 levels of software interrupts.

Isolation Voltage—The voltage that an isolated circuit can normally withstand, usually specified from input to input and/or from any input to the amplifier output, or to the computer bus.

K—When referring to the memory capacity of a computer, it is equal to 2^{10} or 1024 in decimal numbers.

Linearity—A measure for departure from a straight line response in the relationship between two quantities, where the change in one quantity is directly proportional to a change in the other quantity.

LSB—Lease Significant Bit.

MSB—Most Significant Bit.

Machine Language—Instructions that are written in a binary or hexadecimal format that your computer can execute directly. Also: object code, object language.

Microprocessor—The small central processing unit (CPU) that performs the logic operations in a microcomputer system. The CPU also decodes instructions from a stored program, performs arithmetic logic operations, generates timing signals, produces commands for external use in process control, instrumentation and data acquisition hardware control.

Microstepping—The ability to divide a full step of a stepping motor into smaller increments. Microstepping Drivers divide each full step into 10 to 256 μsteps.

Multiplexer—A set of semiconductor or electromechanical switches with a common output that can select one of a number of input signals. With the addition of multiplexing panels, 64, 256, more inputs can be fed to a single 16-channel board. This results in a slower sample rate (throughput), but allows very large data acquisition systems to be constructed affordably.

Multiasking—A property of an operating system in which several processes can be run simultaneously.

Noise—An undesirable electrical signal, noise comes from external sources such as AC power lines, motors, generators, transformers, fluorescent lights, CRT displays, computers, and from internal sources such as semiconductors, resistors, inductors and capacitors.

NMRR—**N**ormal **M**ode **R**ejection **R**atio—The ability of the board to filter out noise from external sources, such as AC power lines. NMRR filtering compensates for transient changes in the incoming signal to provide greater accuracy. The higher the NMRR, the better the filtering of incoming data will be.

Number of Channels—This is the number or input lines the board can sample. Single-Ended inputs share the same ground connection, while Differential inputs have individual two-wire inputs for each incoming signal, allowing greater accuracy and signal isolation. See also **Multiplexer**.

Nyquist Sampling Theorem—A law of sampling theory that states: if a continuous bandwidth limited signal contains no frequency components higher than half the frequency at which it is sampled, then the original signal can be recovered without distortion.

On-Board Memory—Incoming data is stored in on-board memory before being dumped into the PC's memory. On a high-speed board, data is acquired at a much higher rate than can be written into PC memory, so it is stored in the on-board buffer memory.

Operating System—Base-level software that controls a computer, runs programs, interacts with users, and communicates with installed hardware or peripheral devices (i.e., *MS-DOS, UNIX,* or the *Macintosh OS*).

Optical Isolation—The technique of using an optoelectric transmitter and receiver to transfer data without electrical continuity, to eliminate high potential differences and damaging effects transient power surges.

Pretrigger—Boards with "Pretrigger" capability keep a continuous buffer filled with data, so when the trigger conditions are met, the sample includes the data leading up to the trigger condition. **FAST, Flash, HSDAS, PCLAB, MIO,** and **DAP** series boards all provide this capability.

Program I/O—The standard method of memory access, where each piece of data is assigned to a variable and stored individually by the PC's processor.

Protocol—The exact sequence of bits, characters and control codes used to transfer data between computers and peripherals through a communications channel, such as GPIB (IEEE-488).

Range Select—The full-scale range the board uses is selected by one of three methods: through the DAS software, by a hardware jumper on the board, or through the use of an external reference voltage.

Real Time—A property of an event or system in which data is processed as it is acquired instead of being accumulated and processed at a later time.

Reduced Instruction Set Computing (RISC)—A type of microprocessor design that focuses on rapid and efficient processing of a relatively small set of instructions. RISC design is based on the premise that most of the instruction decodes and executes are simple. As a result, the RISC architecture limits the number of instructions that are built into the microprocessor but optimizes each instruction so that it can be carried out very rapidly.

Relay—A switch activated by electricity. A relay allows another signal to be controlled without the need to route the other signal the control point; it also allows a relatively low-power signal—the signal used to activate the relay—to control a high power (voltage) signal.

Resolution—The number of bits in which a digitized value will be stored. This represents the number of divisions into which the full-scale range will be divided (e.g., a 0–10V range with a 12-bit resolution will have 4096 (2^{12}) divisions of 2.44mV each ($10V/2^{12}$ or $10V/4096$)).

RTSI Bus—The **R**eal **T**ime **S**ystem **I**ntegration bus is an additional connector present on some of our DAS boards, allowing you to connect two or more of these boards together. It allows the boards to share data, timing, and interrupt information, at DMA transfer rates up to 2.4 megabytes per second, leaving the PC bus free for other bus operations. This allows synchronized data acquisition, faster DMA transfers, and increased overall performance. Our **MIO, NIH, EISA, DIO 32F,** and **DSP 2200** boards feature this bus.

Sample-and-Hold (S/H)—A circuit that acquires and stores an analog voltage on a capacitor for a short period of time.

Self-Calibrating—A self-calibrating board has an extremely stable on-board reference which is used to calibrate A/D and D/A circuits for higher accuracy.

Self-Diagnostics—These boards have an on-board diagnostic routine which tests most, if not all, of the board's functions at power-up or on request.

Simultaneous Sampling—The ability to acquire and store multiple signals at exactly the same moment. Sample-to-sample inaccuracy is typically measured in nanoseconds. Our **PC 30DS4** simultaneously samples 4 signals to within 300 picoseconds (± 0.3 ns).

Single-Ended—See **Number of Channels**.

Software Drivers—Typically a set of programs or subroutines allowing the user to control basic board functions, such as setup and data acquisition. These can be incorporated into user-written programs to create a simple but functional DAS system.

Software Trigger—Indicates the board allows software control of data acquisition triggering. All of our boards are designed for software control.

Speed/Trigger Throughput—The maximum rate at which the board can sample and convert incoming signals. The typical throughput is divided by the number of channels being sampled to arrive at the samples/second on each channel. To avoid false readings, the samples per second on each channel needs to be greater than twice the frequency of the analog signal being measured.

Step—Stepping motors allow precise rotational positioning, with each increment of rotation being a step. **ORM 200** series motors offers 200 steps per revolution, with each step at 1.8° movement. Half-stepping allows your motor to move in half-step increments (typically 0.9°/half-step).

Successive Approximation ADC—An analog-to-digital converter that sequentially compares a series of binary-weighted values with an analog input to produce an output digital word in n steps, where n is the resolution in bits, of the converter.

Surge Protector/Suppressor—A device that prevents potentially damaging power surges from reaching a computer on any other device that is connected to it. Surge protectors work by collecting and diffusing excess power, sometimes within a few billionths of a second.

Transducer—Any device which generates an electrical signal from real-world physical measurements (e.g., LVDTs, strain gauges, thermocouples, RTDs, etc.)

24-Hour Time-of-Day-Mode—Counter/Timer circuits which may be used as real-time clocks. They are able to control triggering based on the time of day.

Unipolar Inputs—When set to accept a unipolar signal, the channel detects and converts only positive voltages in the 0-X Volts range. (Example: 0 to +10V).

Uninterruptible Power Supply (UPS)—Commonly referred to as a "Battery Back-Up", a UPS provides power for a short period of time after an AC power failure to allow organized shutdown and saving of your work.

Index